Springer Series in Chemical Physics

Volume 127

The Springer Series in Chemical Physics consists of research monographs in basic and applied chemical physics and related analytical methods. The volumes of this series are written by leading researchers of their fields and communicate in a comprehensive way both the basics and cutting-edge new developments. This series aims to serve all research scientists, engineers and graduate students who seek up-to-date reference books.

Marzieh Rabiei · Arvydas Palevicius ·
Sohrab Nasiri · Giedrius Janusas

Hydroxyapatite-Based Nanocomposites

Structure, Mechanics and New Methods

Springer

Marzieh Rabiei
Department of Mechanical Engineering
Faculty of Mechanical Engineering
and Design
Kaunas University of Technology
Kaunas, Lithuania

Arvydas Palevicius
Department of Mechanical Engineering
Faculty of Mechanical Engineering
and Design
Kaunas University of Technology
Kaunas, Lithuania

Sohrab Nasiri
Department of Mechanical Engineering
Faculty of Mechanical Engineering
and Design
Kaunas University of Technology
Kaunas, Lithuania

Giedrius Janusas
Department of Mechanical Engineering
Faculty of Mechanical Engineering
and Design
Kaunas University of Technology
Kaunas, Lithuania

ISSN 0172-6218 ISSN 2364-9003 (electronic)
Springer Series in Chemical Physics
ISBN 978-3-032-12582-8 ISBN 978-3-032-12583-5 (eBook)
https://doi.org/10.1007/978-3-032-12583-5

This Springer imprint is published by the registered company Springer Nature Switzerland AG
The registered company address is: Gewerbestrasse 11, 6330 Cham, Switzerland

Preface

This book presents an innovative approach to the synthesis and mechanical characterization of hydroxyapatite (HA)-based nanocomposites for biotechnological applications, based on the authors' latest scientific papers. By integrating advanced X-ray diffraction (XRD) techniques with ultrasonic pulse-echo testing, it provides a high-precision method for determining nanocrystal size, stress-strain behavior, and elastic modulus. The book investigates the effects of doping HA with silver and copper iodide, enhancing structural integrity and bioactivity, and offering new perspectives for optimizing HA-based materials in biomedical applications. In addition, one PhD thesis was defended in this field by the first co-author.

The book explores the investigation of nanocrystal size of natural HA using X-ray diffraction, as well as the evaluation of an innovative technique for measuring the modulus of elasticity related to the atomic density of planes in unit cell and super cells of crystal lattices. It also examines the relationship between Young's modulus and planar density in unit cells, super cells ($2\times2\times2$), and symmetry cells of cubic crystal lattices. Furthermore, the book explores the effect of calcination temperature on the mechanical and photophysical properties of CuI-doped HA, providing a deeper understanding of material stability under varying conditions.

By linking fundamental materials science with applied biomedical engineering, this book establishes a robust framework for the development of next-generation biomaterials. The combination of innovative synthesis techniques and advanced mechanical characterization offers practical insights to improve the longevity and performance of HA-based implants and scaffolds.

This book will be a valuable resource for researchers, engineers, and professionals in materials science, nanotechnology, and biomedical engineering. It will particularly benefit those working with bioactive materials, implant development, and mechanical characterization, as it provides cutting-edge methods for optimizing HA composites for clinical use. With its systematic presentation of theory, experimental methods, and practical case studies, this book is well suited as a textbook for advanced graduate and PhD courses in materials science, microsystems engineering, nanotechnology, and biomaterials engineering.

Kaunas, Lithuania
2026

Arvydas Palevicius
Giedrius Janusas
Marzieh Rabiei
Sohrab Nasiri

Competing Interests The authors have no competing interests to declare that are relevant to the content of this manuscript.

Contents

Chapter 1
Introduction

This monograph introduces a novel method for the synthesis and mechanical characterization of hydroxyapatite (HA)-based nanocomposites intended for biotechnological use.

There is a necessity to substitute bone lost to both traumatic and non-traumatic incidents. This can be achieved using either endogenous or exogenous bone tissue, but each method presents certain challenges. Employing endogenous bone involves another surgical procedure [1]. Furthermore, the body's native bone supply is limited. The primary drawbacks of exogenous bone implants include potential rejection by the body, the risk of disease transmission with the implant, and the inferior clinical performance compared to fresh endogenous graft materials [2]. Consequently, there is an increasing interest in developing artificial hard tissue replacement implants. Annually, the biomaterials market produces USD 2.3 billion in global sales specifically for hard tissue repair and replacement, within a total market worth USD 12 billion [2]. It is anticipated that the use of biomaterials in clinical treatments will increase by 7–12% per year. Despite this growth, the required quantity of these materials is expected to remain relatively low—only a few tens of tons—unlike in other emerging technical sectors where thousands of tons are used. Metals are commonly employed in critical load-bearing orthopedic applications. However, metallic elements in the body often face challenges like corrosion, wear, and negative tissue responses [3]. Consequently, most metal implants get encapsulated in a dense fibrous layer, impairing optimal load transfer and potentially causing the implant to loosen. As a result, a variety of ceramic materials have been utilized in clinical settings [4]. Notably, ZrO2 and Al2O3 are recognized for their strong mechanical properties and favorable biocompatibility, yet, like metals, they are classified as bioinert materials. According to Hench, tissue responses to implants can be categorized in the following ways [4Tissues react differently to materials based on their properties: toxic materials cause tissue death; non-toxic, biologically inactive (bioinert) materials lead to the formation of fibrous tissue with variable thickness; non-toxic, biologically active (bioactive) materials form an interfacial bond. Notably, calcium phosphates and bioactive glasses demonstrate both

M. Rabiei et al., *Hydroxyapatite-Based Nanocomposites*, Springer Series in Chemical Physics 127, https://doi.org/10.1007/978-3-032-12583-5_1

high bioactivity and biocompatibility. In general, biocompatibility indicates that an implant is well-tolerated on tissue surfaces and encompasses characteristics such as non-toxicity, non-carcinogenicity, chemical inertness, and material stability within the body. These phenomena are discussed in several reviews, such as [5, 6]. Nonetheless, their limited mechanical strength confines their use to small, non-weight bearing implants, powders, coatings, and slightly loaded porous implants [7]. Globally, Ti-alloy hip prostheses with ceramic heads (aluminum oxide or zirconium dioxide) are in use, with about 500,000 such procedures conducted so far, rising by 100,000 annually. In the UK, out of 40,000 annual hip replacements, 18 are revisions [4]. Issues arise due to the loosening of implants stemming from their bioinert qualities and/or stress concentration related to the implant's greater stiffness compared to natural bone. Consequently, there is a pressing need to develop "second generation" bioactive implants that encourage the regeneration of surrounding tissues. Such materials could be applicable for hip replacements as well as other types of artificial bones or dental implants. For an implant to succeed clinically, it must create a stable interface with connective tissue and ensure that its mechanical properties align with those of the tissue it replaces. Hard tissue replacement implants should be bioactive (establishing a chemical bond at the bone/implant interface), possess a modulus akin to bone, and exceed bone in hardness [8]. In cases of trauma, the bone should fracture instead of the implant since implants do not naturally heal like bone and their removal from the body is highly challenging [8]. To develop an ideal implant material, one must meet the necessary requirements for sufficient strength; afterward, the material should gradually biodegrade and be replaced by natural host tissue. Bone's organic elements, primarily collagen, act as a compliant material possessing high toughness, a low modulus, and other attributes typical of polymers. The inorganic components, specifically HA crystals, impart the necessary stiffness to the bone. As a composite of ceramic and organic matter, bone features both high toughness and a relatively elevated modulus. This toughness is attributed not only to collagen but also to its complex fibrous microstructure. It's important to recognize that bone performs as a tough material under low strain rates but fractures more like a brittle material under high strain rates [9]. The bone's remarkable toughness at low strain rates is largely due to its hierarchical structure, which limits crack propagation. Key toughening mechanisms include microcracks in the plastic region of the stress–strain curve, as well as crack deflection and pull-out effects [2].

Bone's mechanical properties are significantly influenced by factors such as moisture content, loading type, loading direction, and bone type. An increase in the bone's mineralization enhances its strength but reduces its fracture elongation. Furthermore, bone's mechanical properties are affected by the orientation of collagen fibers, bone density, porosity, and the molecular structure and arrangement of apatite crystals within the collagen matrix. Additionally, both bone strength and volume significantly decline with age [10]. Various methods for producing HA powder have been extensively covered in literature [11–14]. The production of HA powders primarily involves two main methods: wet processes and solid-state reactions. Wet processes [2, 15] for HA production are subdivided into three categories, including hydrothermal precipitation processes [16] and the hydrolysis of other calcium phosphates [17].

Different processes yield materials with varying morphologies, stoichiometry, and crystallinity. On the other hand, solid-state reactions [18] typically produce a stoichiometric and well-crystallized product, albeit requiring relatively high temperatures and extended heat treatment durations. The sintering capability of these powders is generally limited. By employing precipitation methods at temperatures below 100 °C, nanoparticles can be formed, which manifest as sheets, needles, rods, or equated particles. The crystallinity and Ca/P ratio of these particles are significantly influenced by the specific production conditions and are often less than those found in well-crystallized stoichiometric hydroxyapatite. The hydrothermal method typically yields HA materials with high crystallinity and a Ca/P ratio approaching the stoichiometric balance, with crystal sizes varying from nano to millimeter scales. Hydrolyzing tricalcium phosphate, monetite, brushite, or octacalcium phosphate at low temperatures (typically under 100 °C) produces HA in the form of needles or sheets in the micrometer range [19]. In recent years, synthetic hydroxyapatite has emerged as a crucial material for bone substitutes in both orthopedics and dentistry due to its chemistry and biological compatibility with the mineral components of human bone [20]. Notably, synthetic hydroxyapatite exhibits bioactivity; it is particularly known for its ability to establish a chemical bond with surrounding hard tissues upon implantation (e.g., [4, 21]). In general, synthetic apatites are produced using high-temperature techniques like sintering, leading to highly crystalline structures that demonstrate minimal or no bioresorption activity. This contrasts with nanocrystalline or biocrystalline apatites, which tend to be non-stoichiometric with significantly increased bioactivity. Lately, synthesizing HA via sol–gel processes has become a notable research focus [22]. The sol–gel method primarily offers the benefit of forming and merging apatite crystals at lower temperatures compared to traditional methods. For instance, sintering small apatite crystals created by wet precipitation typically demands temperatures exceeding 1000 °C, whereas sol–gel HA can be densified at significantly lower temperatures [23]. Moreover, sol–gel processed HA often results in a fine microstructure with a blend of nano to submicron-sized crystals, facilitating better integration with host tissue. A new water-based sol–gel technique has been introduced for synthesizing HA at lower temperatures using triethyl phosphite and calcium nitrate as the phosphorous and calcium sources, respectively [22]. It has been demonstrated that the formation of apatite can occur at temperatures down to 350 °C. This method, when applied as coatings, produces a dense, fine-grained apatite layer that adheres strongly to a titanium alloy substrate after being consolidated in air at temperatures below 400 °C. HA synthesis followed a two-step procedure involving an initial 24-h hydrolysis of the phosphate with water, followed by the addition of an aqueous nitrate solution. Schrotter et al. [24] recently described the structural transformations of the phosphate precursor ($P(OC_2H_5)_3$) during hydrolysis, noting that the phosphite quickly reacted with water to yield ($HPO(OC_2H_5)_2$) and subsequently a mix of $HPO(OC_2H_5)(OH)$ and $HPO(OH)_2$. It took several days for the reaction to complete, resulting in the formation of $HPO(OH)_2$.

The sol–gel technique has been gaining significant attention recently due to its renowned benefits in the fabrication of glass, glass–ceramic, and ceramic powders. These benefits include even molecular distribution, low-temperature processing, the capability to produce various particle sizes, and tremendous adaptability for creating nanocrystalline powders, amorphous monoliths, and thin films. It is particularly effective for surface coating, facilitating the production of superior HA thin films on metal substrates [25]. Consequently, the sol–gel process is advantageous for synthesizing both HA powders and films under very gentle conditions. Its versatility enables the creation of thin films through a straightforward process, offering a viable alternative to the commonly employed thermal spraying in biomedical applications [26]. The sol–gel-derived product is characterized by ultra-small primary particles, which significantly enhances contact response and interface stability between synthetic and natural bone. Additionally, the high reactivity of sol–gel powder allows for a reduction in processing temperature and minimizes degradation during sintering [22]. However, a drawback is the possible phosphate hydrolysis and the expensive raw materials [22]. The method proposed here addresses these issues. Yet, many sol–gel processes demand strict pH regulation, intense stirring, and extended hydrolysis time. The presented work describes a simple gel preparation method using a non-alkoxide-based sol–gel process with cost-effective starting materials. Remarkably, gel formation was achieved without a catalyst, providing much gentler conditions for HA film synthesis. This significantly enhances structural integrity while largely avoiding defects associated with plasma spraying [27]. Furthermore, synthesizing at reduced temperatures is particularly advantageous for metal substrates, as it prevents mechanical degradation or phase transitions in the underlying titanium or titanium alloy (such as the $\alpha \rightarrow \beta$ phase transition occurring at 883 and 960 °C, respectively). Nonetheless, treating HA sol–gel films thermally under vacuum is often necessary to avert metal oxidation, yet this process may induce structural instability in the HA coating (e.g., formation of structural water under vacuum) during heat treatment. Thus, it is economically and practically favorable to conduct thermal treatment of the HA coating in air at temperatures lower than the substrate's transition temperature. To reduce substrate oxidation, the thermal treatment temperature should be minimized while ensuring the HA film retains high crystallinity, film integrity, and strong adhesion to the substrate. At temperatures less than approximately 500 °C, oxidation of the Ti or titanium alloy substrate is negligible due to the naturally dense oxide layer present in commercial titanium implants [28]. As a result, the aim of this preliminary study is to identify a sol–gel process capable of producing a high-quality HA coating following heat treatment at temperatures under 500 °C.

Nanocrystals of HA measure approximately $4 \times 50 \times 50$ nm. These minerals are attached to collagen through non-collagenous proteins like osteocalcin, osteopontin, and osteonectin, which constitute roughly 3–5% of bone and serve as key locations for biomineralization and cell adhesion [29]. The chemical composition and behavior of natural HA can differ from that of synthetic HA. Conversely, human bones do not contain pure or stoichiometric HA. Human bones contain other ions, mainly and in traces Na^+, Mg^{2+}, Fe^{2+}, Cl^-, F^-. In bone, the Ca/P molar ratio is below 1.67, differing from synthetic HA [30]. Calcium-deficient HA attracts biological interest as the mineral part of hard tissues mostly comprises of carbonate-substituted calcium-deficient HA [31], with a Ca/P ratio around 1.5. This composition bears a chemical resemblance to tricalcium phosphate but mirrors the structure of stoichiometric hydroxyapatite [32]. Skeletal diseases, infections, and trauma are significant causes of bone-related disorders, often resulting in irreversible damage and presenting a challenge in clinical treatment [33]. Typically, bone repair materials include polymers, inorganic substances, metals, and their composites [34]. Within the polymeric category, poly(lactic acid) (PLA) [35], polycaprolactone (PCL) [36], poly(lactide–co–glycolide) (PLGA) [37], and poly(l–lactic acid) (PLLA) [38] composites have gained attention due to their exceptional biodegradability and biocompatibility. These polymers are non-toxic, FDA-approved, and can be metabolized by the body without leaving residues, irritation, or toxic effects on tissues. Their degradation rate and mechanical properties can be tailored by varying molecular weight, polymerization, and molding techniques to meet different clinical needs. Nonetheless, some polymers suffer from drawbacks like low mechanical strength, lack of bioactivity, and their acidic degradation products do not enhance cellular and tissue growth [33]. To address the aforementioned challenges, incorporating bioceramic elements into polyester can greatly enhance its mechanical properties and bioactivity. This addition also helps manage the acidic environment arising from polyester degradation, which supports bone growth and mitigates inflammation [39]. Recently, intensive studies have focused on enhancing HA coatings through various additives to increase their compactness, purity, uniformity, and adhesion to substrates. Zhiwei Zhou et al. developed a polydopamine/HA hybrid coating on AZ31 Mg alloy using a hydrothermal technique [40]. Their method resulted in a robust, dense inner layer of polydopamine with an outer layer of hydroxyapatite. Polydopamine improved the anticorrosion properties due to its catechol functional groups, binding to metal ions and absorption of Ca^{2+} ions. Zomorodian et al. investigated the corrosion and cell proliferation properties of a composite coating consisting of a polycaprolactone matrix modified with nano-HA and applied to AZ31-Mg alloys [41]. The Mg alloy underwent pre-treatment using hydrofluoric acid. Due to the polycaprolactone's inability to bond with the Mg substrate, an additional thin layer of polyetherimide was applied on the Mg surface to enhance adhesion.

Due to the presence of calcium phosphates (CaPs) in the human body, modern researchers are exploring their use in the restoration and repair of damaged bones. Among the numerous CaP groups, hydroxyapatite (HA) is particularly prominent and well-studied [42, 43]. A key characteristic of the HA structure is its capacity for numerous substitutions without altering the fundamental crystallographic framework. Figure 1.1 illustrates a diagram of the hexagonal HA unit cell and provides a CIF file of HA prepared through heat treatment. In the structure, calcium ions are found in two distinct positions, with 18 ions compactly arranged to create the hexagonal framework. Each corner of the hexagon houses a calcium ion, surrounded by 12 calcium ions, sharing three hexagons. The spaces between hexagons accommodate three tetrahedral phosphate groups per unit cell. Thanks to the flexible nature of this crystal structure, HA allows for ion substitutions with biologically beneficial ions, often termed as doping. Moreover, it is feasible to substitute calcium, phosphate, and/or hydroxyl ions [44]. A unique aspect of HA is the formation of internal channels by OH- ions along the c-axis, a property that significantly influences its mechanical and physical characteristics [45].

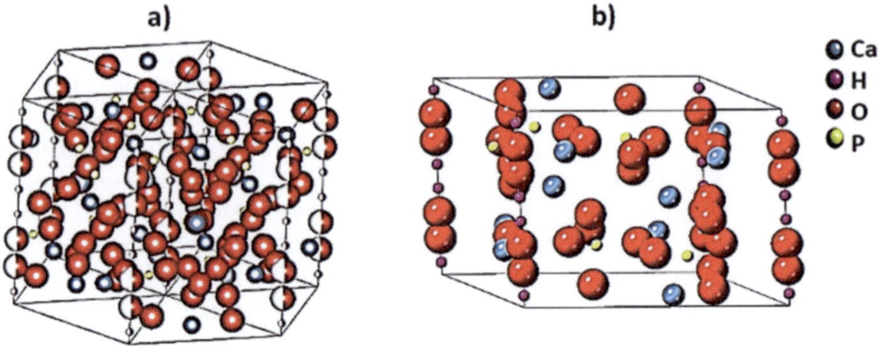

Fig. 1.1 Schematic representation of the **a** HA unit cell and **b** HA structure, extracted from the cif file [8]

One of the primary challenges with employing HA as a bioactive ingredient is the potential for infection. Consequently, the incorporation of antibacterial substances is the optimal solution to this issue [46]. Research has demonstrated that silver (Ag) exhibits antibacterial characteristics [47, 48]. Furthermore, Ag stands out as one of the top biocompatible elements due to its high antibacterial coefficient (100%) [49, 50]. Nonetheless, the biocompatibility of Ag largely hinges on the strength, density, and final phase of the fabricated bioactive composites. Ag is involved in various mechanisms when interacting with biological macromolecules [51]. Additionally, Ag^{+} ions have been found to bind with protein functional groups, causing protein denaturation [52].

Polyvinyltrimethoxysilane (PVTMS) possesses a silanol-like functional group (Si–O-H) that facilitates binding, making it effective against composite degradation [53]. As a polysiloxane-based material, PVTMS is noteworthy for its dense siloxane structure that is interlinked with polymer groups. Additionally, PVTMS exhibits bioactivity and enhances mechanical performance through the formation of a robust Si–O–Si network [54].

The study of copper iodide (CuI) is very interesting as it has several advantages, such as a wide, large band gap, photosensitivity, diamagnetic behaviour, as a dye in solar cells and as a superionic conductor [55, 56]. Several metals, e.g. Ag^{+}, Sr^{2+}, Mn^{2+} and Fe^{3+}, with different weight percentages were used for doping HA by different methods, and the evaluations of the properties were presented in [57–60]. In addition, the Cu^{2+}-doped HA was prepared by the sol–gel method in the study of Renaudin et al. [61]. According to the experimental observations, it was undeniable that the Cu^{2+}-doped HA structure leads to bonding and Cu^{2+} cations are placed along the crystallographic hydroxyl sites by the formation of linear O–Cu–O units. Subsequent studies with copper have shown significant similarities and revealed specific behaviors. In addition, a higher degree of doping can be achieved by a reduction of Cu^{2+} to Cu^{+}, but this requires a higher sintering temperature [62–64].

Uwe Holzwarth and Neil Gibson stated that the Scherrer equation is connected to a distinct X-ray diffraction peak. This equation is denoted with the subscript (hkl) because it pertains to only one particular peak. Notably, the equation is applicable for determining average sizes up to approximately 100 nm. The noise criteria are influenced by the instrumentation and the signal-to-sample ratio, as diffraction peak broadening lessens with greater crystallite size. Differentiating broadening due to crystallite size from other factors can be quite challenging. There are always potential errors, and effective calculation methods are those that minimize these errors to achieve more precise data. Deriving the size of nanoparticles from XRD patterns is infeasible since a particle contains multiple crystals within the nano or micro scale. An X-ray beam penetrates to the crystal size for information, so the size calculation refers to crystals rather than particles. The Scherrer equation refers to the diffraction peak given in Eq. 1.1 [65], where L is the size of the nanocrystal, K is the shape factor, which is usually assumed to be 0.89 for ceramic materials, λ is the wavelength of the radiation in nanometers ($\lambda_{Cu_{K\alpha}} = 0.15405\,nm$), Θ is the diffraction angle of the pea, and β is the full width at half maximum (FWHM) of the peak in radians. In addition, the broadening of the peaks is related to the physical broadening and the instrumental broadening [66, 67].

$$L = \frac{K\lambda}{\beta} \cdot \frac{1}{\cos\theta}\,[67] \tag{1.1}$$

In order to decrease this error of the instrument, Eq. 1.2 can be used:

$$\beta_d^2 = \beta_m^2 - \beta_i^2\,[67]. \tag{1.2}$$

In this formula, β_m is the measured broadening, β_i is the instrumental broadening and β_d is introduced as the corrected broadening, which is responsible for the crystal size. The determination of crystal size, using the Scherrer equation as outlined in [68, 69], can be achieved by analyzing both the instrumental and physical broadening of the sample at its full width at half maximum (FWHM) and adjusting for the physical broadening. Many studies apply the Scherrer equation only to the sharpest peak, neglecting a comprehensive analysis across all peaks. Equation 1.3 presents Hooke's Law, which states that for slight displacements, stress is directly proportional to strain. This fundamental equation serves as the basis for transforming the symmetry into the six configurations of σ and ε [70].

$$\begin{pmatrix} \sigma_{xx} \\ \sigma_{yy} \\ \sigma_{zz} \\ \sigma_{yz} \\ \sigma_{zx} \\ \sigma_{xy} \end{pmatrix} = \begin{pmatrix} c_{11} & c_{12} & c_{13} & c_{14} & c_{15} & c_{16} \\ c_{21} & c_{22} & c_{23} & c_{24} & c_{25} & c_{26} \\ c_{31} & c_{32} & c_{33} & c_{34} & c_{35} & c_{36} \\ c_{41} & c_{42} & c_{43} & c_{44} & c_{45} & c_{46} \\ c_{51} & c_{52} & c_{53} & c_{54} & c_{55} & c_{56} \\ c_{61} & c_{62} & c_{63} & c_{64} & c_{65} & c_{66} \end{pmatrix} = \begin{pmatrix} \varepsilon_{xx} \\ \varepsilon_{yy} \\ \varepsilon_{zz} \\ \varepsilon_{yz} \\ \varepsilon_{zx} \\ \varepsilon_{xy} \end{pmatrix} \tag{1.3}$$

In addition, Hooke's Law can be written as follows (Eq. (1.4)):

$$\begin{aligned}
\sigma_{xx} &= C_{11}\varepsilon_{xx} + C_{12}\varepsilon_{yy} + C_{13}\varepsilon_{zz} + C_{14}\varepsilon_{yz} + C_{15}\varepsilon_{zx} + C_{16}\varepsilon_{xy} \\
\sigma_{yy} &= C_{21}\varepsilon_{xx} + C_{22}\varepsilon_{yy} + C_{23}\varepsilon_{zz} + C_{24}\varepsilon_{yz} + C_{25}\varepsilon_{zx} + C_{26}\varepsilon_{xy} \\
\sigma_{zz} &= C_{31}\varepsilon_{xx} + C_{32}\varepsilon_{yy} + C_{33}\varepsilon_{zz} + C_{34}\varepsilon_{yz} + C_{35}\varepsilon_{zx} + C_{36} \\
\sigma_{yz} &= C_{41}\varepsilon_{xx} + C_{42}\varepsilon_{yy} + C_{43}\varepsilon_{zz} + C_{44}\varepsilon_{yz} + C_{45}\varepsilon_{zx} + C_{46}\varepsilon_{xy} \\
\sigma_{zx} &= C_{51}\varepsilon_{xx} + C_{52}\varepsilon_{yy} + C_{53}\varepsilon_{zz} + C_{54}\varepsilon_{yz} + C_{55}\varepsilon_{zx} + C_{56}\varepsilon_{xy} \\
\sigma_{xy} &= C_{61}\varepsilon_{xx} + C_{62}\varepsilon_{yy} + C_{63}\varepsilon_{zz} + C_{64}\varepsilon_{yz} + C_{65}\varepsilon_{zx} + C_{66}\varepsilon_{xy}
\end{aligned} \tag{1.4}$$

The elastic stiffness determines the response of the crystal to an externally applied stress or strain and provides information about the bonding properties as well as the mechanical and structural stability [71]. The HA system has five elastic constants (Eq. (1.5)). Therefore, the values of the five independent (elastic stiffness constants C_{ij}) can be referred to as C_{11}, C_{12}, C_{13}, C_{33}, C_{44}.

$$\text{Hydroxyapatite matrix}\begin{vmatrix} C_{11} & C_{12} & C_{13} & 0 & 0 & 0 \\ C_{12} & C_{11} & C_{13} & 0 & 0 & 0 \\ C_{13} & C_{13} & C_{33} & 0 & 0 & 0 \\ 0 & 0 & 0 & C_{44} & 0 & 0 \\ 0 & 0 & 0 & 0 & C_{44} & 0 \\ 0 & 0 & 0 & 0 & 0 & \frac{1}{2}(C_{11} - C_{12}) \end{vmatrix} \tag{1.5}$$

For conventional hexagonal systems, such as HA, the relationship between C_{ij} and elastic compliances S_{ij} is introduced in Eqs. (1.6–1.10) [72].

$$S_{11} = \frac{1}{2}\left(\frac{C_{33}}{C_{33}(C_{11} + C_{12}) - 2(C_{13})^2} + \frac{1}{C_{11} - C_{12}}\right) \tag{1.6}$$

$$S_{12} = \frac{1}{2}\left(\frac{C_{33}}{C_{33}(C_{11} + C_{12}) - 2(C_{13})^2} - \frac{1}{C_{11} - C_{12}}\right) \tag{1.7}$$

$$S_{33} = \frac{C_{11} + C_{12}}{C_{33}(C_{11} + C_{12}) - 2(C_{13})^2} \tag{1.8}$$

$$S_{13} = -\frac{C_{13}}{C_{33}(C_{11} + C_{12}) - 2(C_{13})^2} \tag{1.9}$$

$$S_{44} = \frac{1}{C_{44}} \tag{1.10}$$

According to the Eqs. (1.6–1.10), in order to obtain S values, C values are needed. The complete set of five values: C_{11}, C_{12}, C_{13}, C_{33}, and C_{44} of samples, was found from ultrasonic measurements of the phase velocity anisotropy. In these equations, ρ and V are the density of the sample and velocity [73–75].

$$C_{11} = \rho V_{1/1}^2, C_{22} = \rho V_{2/2}^2 \tag{1.11}$$

$$C_{66} = \rho V_{1/2}^2 = \rho V_{2/1}^2, C_{55} = \rho V_{1/3}^2 = \rho V_{3/1}^2 \quad (1.12)$$

$$C_{12} = \sqrt{\left(C_{11} + C_{66} - 2\rho V_{12/12}^2\right)\left(C_{22} + C_{66} - 2\rho V_{12/12}^2\right)} - C_{66} \quad (1.13)$$

$$C_{44} = \rho V_{2/3}^2 = \rho V_{3/2}^2 \quad (1.14)$$

$$C_{13} = \sqrt{\left(C_{11} + C_{55} - 2\rho V_{13/13}^2\right)\left(C_{33} + C_{55} - 2\rho V_{13/13}^2\right)} - C_{55} \quad (1.15)$$

$$C_{33} = \rho V_{3/3}^2 \quad (1.16)$$

The velocities were measured using the standard ultrasonic pulse-echo method ASTM E797/E797-M-15, as referenced in [76]. This method suggests that the shear modulus is related to both the Burgers vector and the modulus of elasticity, while also indicating that the dislocation density aligns with the modulus of elasticity [77, 78]. To determine the elastic modulus, longitudinal and transverse waves were utilized in the ultrasonic technique [79, 80]. Following this approach (Eq. (1.17)), the elastic modulus was calculated from the ultrasonic wave velocities and the sample density.

$$E = \frac{\rho c_1^2\left[3\left(\frac{c_l}{c_t}\right)^2 - 4\right]}{\left(\frac{c_l}{c_t}\right)^2 - 1} \quad (1.17)$$

In Eq. 1.17, ρ, c_l, and c_t are density, velocity of longitudinal and transverse ultrasound waves tandemly. Furthermore, according to Eq. 1.18, the velocity of the longitudinal and transverse waves can be registered in the signals by determining the length of the sample and the differences between two echoes ($t_2 - t_1$) [81].

$$C = \frac{2L}{t}; \quad (1.18)$$

where L is the length of the sample and t is the difference between two echoes, and the density of the sample can be determined by measuring the mass and volume of the sample [36]. In addition, by substituting Eqs. 1.17 and 1.18, the main equation for calculating the modulus of elasticity is introduced as Eq. 1.19 [82].

$$E = \frac{4\rho\left(\frac{L}{t_s}\right)^2\left(3t_s^2 - 4t_1^2\right)}{t_s^2 - t_1^2} \quad (1.19)$$

This monograph introduces a novel method for the synthesis and mechanical characterization of hydroxyapatite (HA)-based nanocomposites intended for biotechnological use. By combining advanced X-ray diffraction (XRD) methods with ultrasonic pulse-echo testing, this research offers a precise technique for evaluating nanocrystal size, stress–strain behaviour, and elastic modulus. The study explores the impact of doping HA with silver and copper iodide to enhance both structural integrity and bioactivity. These findings present a fresh perspective on optimizing HA-based materials for biomedical purposes [83].

References

1. J.R. Parsons, J.L. Ricci, H. Alexander, P.K. Bajpai, Osteoconductive composite grouts for orthopedic use. Ann. N. Y. Acad. Sci. **523**, 190–207 (1988). https://doi.org/10.1111/J.1749-6632.1988.TB38512.X
2. W. Suchanek, M. Yoshimura, Processing and properties of hydroxyapatite-based biomaterials for use as hard tissue replacement implants, J. Mater. Res. 1998 131. **13** (2011) 94–117. https://doi.org/10.1557/JMR.1998.0015.
3. L.L. Hench, E.C. Ethridge, Biomaterials—the interfacial problem. Adv. Biomed. Eng. **5**, 35–150 (1975). https://doi.org/10.1016/B978-0-12-004905-9.50007-4
4. L.L. Hench, Bioceramics: from concept to clinic. J. Am. Ceram. Soc. **74**, 1487–1510 (1991). https://doi.org/10.1111/J.1151-2916.1991.TB07132.X
5. D.F. Williams, Review Tissue-biomaterial interactions. J. Mater. Sci. **22**, 3421–3445 (1987)
6. A. Remes, D.F. Williams, Immune response in biocompatibility. Biomater. **13**, 731–743 (1992). https://doi.org/10.1016/0142-9612(92)90010-L
7. C.M. Botelho, R. Brooks, M. Kanitakahara, C. Ohtsuki, S. Best, M.A. Lopers, N. Rushton, W. Bonfield, J.D. Santos, C.M. Botelho, R. Brooks, M. Kanitakahara, C. Ohtsuki, S. Best, M.A. Lopers, N. Rushton, W. Bonfield, J.D. Santos, Effect of protein adsorption onto the dissolution of silicon-substituted hydroxyapatite. J. Encapsulation Adsorpt. Sci. **1**, 72–79 (2011). https://doi.org/10.4236/JEAS.2011.14010
8. W. Bonfield, M. Wang, K.E. Tanner, Interfaces in analogue biomaterials. Acta Mater. **46**, 2509–2518 (1998). https://doi.org/10.1016/S1359-6454(98)80035-9
9. W.W. Wright, International encyclopedia of composites: Volumes 1 & 2 Edited by Stuart M. Lee, VCH Publishers, New York, 1990. Price DM 450 per Volume. Vol. 1 pp. xvi + 563. ISBN 0–89573–731–0 Vol. 2 pp. xii + 524. ISBN 0–89573–732–9, Polym. Int. 25 (1991) 132–132. https://doi.org/10.1002/PI.4990250219
10. W.J. Landis, J.J. Librizzi, M.G. Dunn, F.H. Silver, A study of the relationship between mineral content and mechanical properties of turkey gastrocnemius tendon. J. Bone Miner. Res. **10**, 859–867 (1995). https://doi.org/10.1002/JBMR.5650100606
11. R. Marzouki, A. Guesmi, M.F. Zid, A. Driss, R. Marzouki, A. Guesmi, M.F. Zid, A. Driss, synthesis, crystal structure and electrical properties of a new mixed compound (Na0.71Ag0.29)2CoP2O7, Cryst. Struct. Theory Appl. 1 (2012) 68–73. https://doi.org/10.4236/CSTA.2012.13013
12. D. Luna-Zaragoza, E.T. Romero-Guzmán, L.R. Reyes-Gutiérrez, D. Luna-Zaragoza, E.T. Romero-Guzmán, L.R. Reyes-Gutiérrez, Surface and physicochemical characterization of phosphates vivianite, Fe2(PO4)3 and Hydroxyapatite, Ca5(PO4)3OH. J. Miner. Mater. Charact. Eng. **8**, 591–609 (2009). https://doi.org/10.4236/JMMCE.2009.88052
13. E.B. Jaffe, Abstracts of the literature on synthesis of apatites and some related phosphates, (n.d.)
14. E. Yılmaz, B. Çakıroğlu, A. Gökçe, F. Findik, H.O. Gulsoy, N. Gulsoy, Ö. Mutlu, M. Özacar, Novel hydroxyapatite/graphene oxide/collagen bioactive composite coating on Ti16Nb alloys

by electrodeposition. Mater. Sci. Eng. C **101**, 292–305 (2019). https://doi.org/10.1016/J.MSEC.2019.03.078
15. S. Lazić, Microcrystalline hydroxyapatite formation from alkaline solutions. J. Cryst. Growth **147**, 147–154 (1995). https://doi.org/10.1016/0022-0248(94)00587-7
16. K. Onuma, A. Ito, T. Tateishi, Investigation of a growth unit of hydroxyapatite crystal from the measurements of step kinetics. J. Cryst. Growth **167**, 773–776 (1996). https://doi.org/10.1016/0022-0248(96)00421-6
17. N.S. Chickerur, M.S. Tung, W.E. Brown, A mechanism for incorporation of carbonate into apatite, Calcif. Tissue Int. 1980 321. **32** (1980) 55–62. https://doi.org/10.1007/BF02408521
18. Y. Ota, T. Iwashita, T. Kasuga, Y. Abe, Novel preparation method of hydroxyapatite fibers. J. Am. Ceram. Soc. **81**, 1665–1668 (1998). https://doi.org/10.1111/J.1151-2916.1998.TB02529.X
19. M. Rabiei, A. Palevicius, A. Monshi, S. Nasiri, A. Vilkauskas, G. Janusas, Comparing methods for calculating nano crystal size of natural hydroxyapatite using X-Ray diffraction, Nanomater. **10**, Page 1627. 10 (2020) 1627. https://doi.org/10.3390/NANO10091627
20. E.E. Berry, The structure and composition of some calcium-deficient apatites. J. Inorg. Nucl. Chem. **29**, 317–327 (1967). https://doi.org/10.1016/0022-1902(67)80033-2
21. P. Ducheyne, L.L. Hench, A. Kagan, M. Martens, A. Bursens, J.C. Mulier, Effect of hydroxyapatite impregnation on skeletal bonding of porous coated implants. J. Biomed. Mater. Res. **14**, 225–237 (1980). https://doi.org/10.1002/JBM.820140305
22. D.M. Liu, T. Troczynski, W.J. Tseng, Water-based sol–gel synthesis of hydroxyapatite: process development. Biomaterials **22**, 1721–1730 (2001). https://doi.org/10.1016/S0142-9612(00)00332-X
23. P. Layrolle, A. Ito, T. Tateishi, Sol-Gel synthesis of amorphous calcium phosphate and sintering into microporous hydroxyapatite bioceramics. J. Am. Ceram. Soc. **81**, 1421–1428 (1998). https://doi.org/10.1111/J.1151-2916.1998.TB02499.X
24. J.-C. Schroti'er, A. Cardenas, M. Smaihi, N. Hovnanian, Silicon and phosphorus alkoxide mixture: sol-gel study by spectroscopic technics, J. Sol-Gel Sci. Technol. 4 (1995) 195–204
25. K. Agrawal, G. Singh, D. Puri, S. Prakash, Synthesis and characterization of hydroxyapatite powder by sol-gel method for biomedical application. J. Miner. Mater. Charact. Eng. **10**, 727–734 (2011). https://doi.org/10.4236/jmmce.2011.108057
26. K. Agrawal, G. Singh, D. Puri, S. Prakash, K. Agrawal, G. Singh, D. Puri, S. Prakash, Synthesis and characterization of hydroxyapatite powder by sol-gel method for biomedical application. J. Miner. Mater. Charact. Eng. **10**, 727–734 (2011). https://doi.org/10.4236/JMMCE.2011.108057
27. C. Chai, B. Ben-Nissan, S. Pyke, L. Evans, Sol-Gel derived hydroxylapatite coatings for biomedical applications, **10** (2007) 205–216. https://doi.org/10.1080/10426919508935016
28. A.P. Ameen, R.D. Short, R. Johns, G. Schwach, The surface analysis of implant materials. The surface composition of a titanium dental implant material., Clin. Oral Implants Res. **4** (1993) 144–150. https://doi.org/10.1034/J.1600-0501.1993.040305.X
29. S. Ramakrishna, J. Mayer, E. Wintermantel, K.W. Leong, Biomedical applications of polymer-composite materials: a review. Compos. Sci. Technol. **61**, 1189–1224 (2001). https://doi.org/10.1016/S0266-3538(00)00241-4
30. R. Murugan, S. Ramakrishna, K. Panduranga Rao, Nanoporous hydroxy-carbonate apatite scaffold made of natural bone, Mater. Lett. **60** (2006) 2844–2847. https://doi.org/10.1016/J.MATLET.2006.01.104.
31. X. Deng, J. Hao, C. Wang, Preparation and mechanical properties of nanocomposites of poly(d,l-lactide) with Ca-deficient hydroxyapatite nanocrystals. Biomater. **22**, 2867–2873 (2001). https://doi.org/10.1016/S0142-9612(01)00031-X
32. S.C. Liou, S.Y. Chen, D.M. Liu, Synthesis and characterization of needlelike apatitic nanocomposite with controlled aspect ratios. Biomater. **24**, 3981–3988 (2003). https://doi.org/10.1016/S0142-9612(03)00303-X
33. Z. Fu, J. Cui, B. Zhao, S.G. Shen, K. Lin, An overview of polyester/hydroxyapatite composites for bone tissue repairing. J. Orthop. Transl. **28**, 118–130 (2021). https://doi.org/10.1016/J.JOT.2021.02.005

34. Q. Pan, Y. Li, J. Xu, Y. Kang, Y. Li, B. Wang, Y.P. Yang, S. Lin, G. Li, The effects of tubular structure on biomaterial aided bone regeneration in distraction osteogenesis. J. Orthop. Transl. **25**, 80–86 (2020). https://doi.org/10.1016/J.JOT.2020.09.009
35. D.K. Patel, S.D. Dutta, J. Hexiu, K. Ganguly, K.T. Lim, Bioactive electrospun nanocomposite scaffolds of poly(lactic acid)/cellulose nanocrystals for bone tissue engineering. Int. J. Biol. Macromol. **162**, 1429–1441 (2020). https://doi.org/10.1016/J.IJBIOMAC.2020.07.246
36. S. Surucu, H. Turkoglu Sasmazel, Development of core-shell coaxially electrospun composite PCL/chitosan scaffolds, Int. J. Biol. Macromol. **92** (2016), 321–328. https://doi.org/10.1016/J.IJBIOMAC.2016.07.013.
37. G. Sen Shi, Y.Y. Li, Y.P. Luo, J.F. Jin, Y.X. Sun, L.Z. Zheng, Y.X. Lai, L. Li, G. hui Fu, L. Qin, S.H. Chen, Bioactive PLGA/tricalcium phosphate scaffolds incorporating phytomolecule icaritin developed for calvarial defect repair in rat model, J. Orthop. Transl. **24** (2020), 112–120. https://doi.org/10.1016/J.JOT.2020.05.008.
38. X. Yang, L. Li, D. Yang, J. Nie, G. Ma, Electrospun Core-Shell Fibrous 2D Scaffold with Biocompatible Poly(Glycerol Sebacate) and Poly-l-Lactic Acid for Wound Healing. Adv. Fiber Mater. **2**, 105–117 (2020). https://doi.org/10.1007/S42765-020-00027-X/FIGURES/6
39. L.L. Hench, Sol-gel materials for bioceramic applications. Curr. Opin. Solid State Mater. Sci. **2**, 604–610 (1997). https://doi.org/10.1016/S1359-0286(97)80053-8
40. Z. Zhou, B. Zheng, Y. Gu, C. Shen, J. Wen, Z. Meng, S. Chen, J. Ou, A. Qin, New approach for improving anticorrosion and biocompatibility of magnesium alloys via polydopamine intermediate layer-induced hydroxyapatite coating. Surf. Interfaces. **19**, 100501 (2020). https://doi.org/10.1016/J.SURFIN.2020.100501
41. A. Zomorodian, M.P. Garcia, T. Moura E Silva, J.C.S. Fernandes, M.H. Fernandes, M.F. Montemor, Biofunctional composite coating architectures based on polycaprolactone and nanohydroxyapatite for controlled corrosion activity and enhanced biocompatibility of magnesium AZ31 alloy, Mater. Sci. Eng. C. **48**(2015) 434–443. https://doi.org/10.1016/J.MSEC.2014.12.027.
42. M. Wang, B. Duan, Materials and their biomedical applications, Encycl. Biomed. Eng. **1–3**, 135–152 (2019). https://doi.org/10.1016/B978-0-12-801238-3.99860-X
43. R. Marzieh, N. Sohrab, P. Arvydas, J. Giedrius, Preparation and investigation of bioactive organic-inorganic nano-composite derived from PVB-co-VA-co-VAc/HA, 15th Int. Conf. Mechatron. Syst. Mater. MSM 2020. (2020). https://doi.org/10.1109/MSM49833.2020.9201632
44. Y. Han, Q. Wei, P. Chang, K. Hu, O.V. Okoro, A. Shavandi, L. Nie, Three-Dimensional Printing of Hydroxyapatite Composites for Biomedical Application, Cryst. Page **11**, 353 (2021). https://doi.org/10.3390/CRYST11040353
45. V. Bystrov, E. Paramonova, L. Avakyan, J. Coutinho, N. Bulina, Simulation and Computer Study of Structures and Physical Properties of Hydroxyapatite with Various Defects, Nanomater. **11** 2752, (2021). https://doi.org/10.3390/NANO11102752
46. V.K. Sharma, R.A. Yngard, Y. Lin, Silver nanoparticles: Green synthesis and their antimicrobial activities. Adv. Colloid Interface Sci. **145**, 83–96 (2009). https://doi.org/10.1016/J.CIS.2008.09.002
47. L. Mei, Z. Lu, X. Zhang, C. Li, Y. Jia, Polymer-Ag nanocomposites with enhanced antimicrobial activity against bacterial infection. ACS Appl. Mater. Interfaces **6**, 15813–15821 (2014). https://doi.org/10.1021/AM502886M
48. R.G. Saratale, G.D. Saratale, G. Ghodake, S.K. Cho, A. Kadam, G. Kumar, B.H. Jeon, D. Pant, A. Bhatnagar, H.S. Shin, Wheat straw extracted lignin in silver nanoparticles synthesis: Expanding its prophecy towards antineoplastic potency and hydrogen peroxide sensing ability. Int. J. Biol. Macromol. **128**, 391–400 (2019). https://doi.org/10.1016/J.IJBIOMAC.2019.01.120
49. L. Badrour, A. Sadel, M. Zahir, L. Kimakh, A. El Hajbi, Synthesis and physical and chemical characterization of Ca10−xAgx(PO4)6(OH)2–xx apatites. Ann. Chim. Sci. Des Matériaux. **23**, 61–64 (1998). https://doi.org/10.1016/S0151-9107(98)80012-3

50. G.D. Saratale, R.G. Saratale, D.S. Kim, D.Y. Kim, H.S. Shin, Exploiting Fruit Waste Grape Pomace for Silver Nanoparticles Synthesis, Assessing Their Antioxidant, Antidiabetic Potential and Antibacterial Activity Against Human Pathogens: A Novel Approach, Nanomater. **10**, 1457 (2020). https://doi.org/10.3390/NANO10081457
51. C.S. Ciobanu, S.L. Iconaru, M.C. Chifiriuc, A. Costescu, P. Le Coustumer, D. Predoi, Synthesis and antimicrobial activity of silver-doped hydroxyapatite nanoparticles, Biomed Res. Int. (2013). https://doi.org/10.1155/2013/916218
52. S.K. Rastogi, V.J. Rutledge, C.L. Gibson, D.A. Newcombe, J.R. Branen, A.L. Branen, Ag colloids and Ag clusters over EDAPTMS-coated silica nanoparticles: synthesis, characterization, and antibacterial activity against Escherichia coli, Nanomedicine Nanotechnology. Biol. Med. **7**, 305–314 (2011). https://doi.org/10.1016/J.NANO.2010.11.003
53. M. Lim, D. Kim, J. Seo, H. Han, Preparation and properties of poly(vinyl alcohol)/vinyltrimethoxysilane (PVA/VTMS) hybrid films with enhanced thermal stability and oxygen barrier properties, Macromol. Res. 2210. 22 (2014) 1096–1103. https://doi.org/10.1007/S13233-014-2146-2
54. A. Motalebi, M. Nasr-Esfahani, R. Ali, M. Pourriahi, Improvement of corrosion performance of 316L stainless steel via PVTMS/henna thin film. Prog. Nat. Sci. Mater. Int. **22**, 392–400 (2012). https://doi.org/10.1016/J.PNSC.2012.10.006
55. M. Ferhat, A. Zaoui, M. Certier, J.P. Dufour, B. Khelifa, Electronic structure of the copper halides CuCl, CuBr and CuI. Mater. Sci. Eng. B **39**, 95–100 (1996). https://doi.org/10.1016/0921-5107(95)01518-3
56. V.N. Ermakov, S.P. Kruchinin, A. Fujiwara, Electronic nanosensors based on nanotransistor with bistability behaviour, 341–349 (2008). https://doi.org/10.1007/978-1-4020-9146-9_26
57. Y. Huang, Z. Xu, X. Zhang, X. Chang, X. Zhang, Y.C. Li, T. Ye, R. Han, S. Han, Y. Gao, X. Du, H. Yang, Nanotube-formed Ti substrates coated with silicate/silver co-doped hydroxyapatite as prospective materials for bone implants. J. Alloys Compd. **697**, 182–199 (2017). https://doi.org/10.1016/j.jallcom.2016.12.139
58. A. Dubnika, D. Loca, V. Rudovica, M.B. Parekh, L. Berzina-Cimdina, Functionalized silver doped hydroxyapatite scaffolds for controlled simultaneous silver ion and drug delivery. Ceram. Int. **43**, 3698–3705 (2017). https://doi.org/10.1016/j.ceramint.2016.11.214
59. C.S. Ciobanu, S.L. Iconaru, F. Massuyeau, L.V. Constantin, A. Costescu, D. Predoi, Synthesis, structure, and luminescent properties of europium-doped hydroxyapatite nanocrystalline powders, J. Nanomater. (2012). https://doi.org/10.1155/2012/942801
60. M. Mirzaee, M. Vaezi, Y. Palizdar, Synthesis and characterization of silver doped hydroxyapatite nanocomposite coatings and evaluation of their antibacterial and corrosion resistance properties in simulated body fluid. Mater. Sci. Eng. C **69**, 675–684 (2016). https://doi.org/10.1016/j.msec.2016.07.057
61. G. Renaudin, S. Gomes, J.-M. Nedelec, First-Row transition metal doping in calcium phosphate bioceramics: a detailed crystallographic study. Materials (Basel). **10**, 92 (2017). https://doi.org/10.3390/ma10010092
62. S. Gomes, A. Kaur, J.M. Grenèche, J.M. Nedelec, G. Renaudin, Atomic scale modeling of iron-doped biphasic calcium phosphate bioceramics. Acta Biomater. **50**, 78–88 (2017). https://doi.org/10.1016/j.actbio.2016.12.011
63. S. Khamseh, M. Ganjaee Sari, E. Alibakhshi, M. Nemati Valandaran, Hydrogen-free Cu: Amorphous-C: N Coating on TC4 Titanium Alloy: The Role of Gas Ratio on Mechanical and Antibacterial Potency, Prog. Color. Color. Coatings. **14** (2021) 281–291. https://doi.org/10.30509/pccc.2021.166770.1102
64. A. V. Artem'ev, M.P. Davydova, X. Hei, M.I. Rakhmanova, D.G. Samsonenko, I.Y. Bagryanskaya, K.A. Brylev, V.P. Fedin, J.-S. Chen, M. Cotlet, J. Li, Family of Robust and Strongly Luminescent CuI-Based Hybrid Networks Made of Ionic and Dative Bonds, Chem. Mater. **32** 10708–10718, (2020). https://doi.org/10.1021/ACS.CHEMMATER.0C03984
65. Nachrichten von der Gesellschaft der Wissenschaften zu Göttingen, Mathematisch-Physikalische Klasse, n.d

66. R. Das, S.S. Nath, R. Bhattacharjee, Preparation of linoleic acid capped gold nanoparticles and their spectra. Phys. E Low-Dimensional Syst. Nanostructures. **43**, 224–227 (2010). https://doi.org/10.1016/J.PHYSE.2010.07.008
67. P.C. Dey, R. Das, Effect of silver doping on the elastic properties of CdS nanoparticles, (n.d.). https://doi.org/10.1007/s12648-018-1214-4
68. R.S. Darweesh, N.M. Ayoub, S. Nazzal, Gold nanoparticles and angiogenesis: molecular mechanisms and biomedical applications. Int. J. Nanomedicine **14**, 7643–7663 (2019). https://doi.org/10.2147/IJN.S223941
69. Y.T. Prabhu, K.V. Rao, V.S.S. Kumar, B.S. Kumari, Y.T. Prabhu, K.V. Rao, V.S.S. Kumar, B.S. Kumari, X-Ray analysis by williamson-hall and size-strain plot methods of ZnO nanoparticles with fuel variation. World J. Nano Sci. Eng. **4**, 21–28 (2014). https://doi.org/10.4236/WJNSE.2014.41004
70. K. Rajabi, S. Hosseini-Hashemi, Application of the generalized Hooke's law for viscoelastic materials (GHVMs) in nanoscale mass sensing applications of viscoelastic nanoplates: A theoretical study. Eur. J. Mech. - A/Solids. **67**, 71–83 (2018). https://doi.org/10.1016/J.EUROMECHSOL.2017.08.015
71. M.B. Kanoun, S. Goumri-Said, K. Abdullah, Theoretical study of physical properties and oxygen incorporation effect in nanolaminated ternary carbides 211-MAX phases, Adv. Sci. Technol. Mn+1AXn Phases. 177–196 (2012). https://doi.org/10.1533/9780857096012.177.
72. Y. Li, R.B. Thompson, Relations between elastic constants Cij and texture parameters for hexagonal materials. J. Appl. Phys. **67**, 2663 (1998). https://doi.org/10.1063/1.345479
73. M. Mah, D.R. Schmitt, Correction to "Determination of the complete elastic stiffnesses from ultrasonic phase velocity measurements." J. Geophys. Res. Solid Earth. **108**, 2490 (2003). https://doi.org/10.1029/2003JB002710
74. J.R. Neighbours, G.E. Schacher, Determination of elastic constants from sound-velocity measurements in crystals of general symmetry. J. Appl. Phys. **38**, 5366 (2004). https://doi.org/10.1063/1.1709328
75. W.C. Van Buskirk, S.C. Cowin, R.N. Ward, Ultrasonic measurement of orthotropic elastic constants of bovine femoral bone. J. Biomech. Eng. **103**, 67–72 (1981). https://doi.org/10.1115/1.3138262
76. ASTM E797/E797M-15—Standard practice for measuring thickness by manual ultrasonic pulse-echo contact method, (n.d.). https://webstore.ansi.org/Standards/ASTM/astme797e797m15 (Accessed May 7, 2022)
77. Mechanics of Microstructured Materials, Mech. Microstruct. Mater. (2004). https://doi.org/10.1007/978-3-7091-2776-6
78. N. Pandech, K. Sarasamak, S. Limpijumnong, Elastic properties of perovskite ATiO3 (A = Be, Mg, Ca, Sr, and Ba) and PbBO3 (B = Ti, Zr, and Hf): First principles calculations. J. Appl. Phys. **117**, 174108 (2015). https://doi.org/10.1063/1.4919837
79. H. Wang, P.L. Prendiville, P.J. McDonnell, W.V. Chang, An ultrasonic technique for the measurement of the elastic moduli of human cornea. J. Biomech. **29**, 1633–1636 (1996). https://doi.org/10.1016/S0021-9290(96)80017-0
80. Nondestructive Evaluation: A Tool in Design, Manufacturing and Service - Don E. Bray, Roderick K. Stanley - Google Books, (n.d.). https://books.google.com.et/books?id=5WtmjwEACAAJ&printsec=copyright#v=onepage&q&f=false (Accessed May 7, 2022)
81. M. Rabiei, A. Palevicius, S. Nasiri, A. Dashti, A. Vilkauskas, G. Janusas, Relationship between Young's Modulus and Planar Density of Unit Cell, Super Cells ($2 \times 2 \times 2$), Symmetry Cells of Perovskite (CaTiO3) Lattice, Mater. 2021, Vol. 14, Page 1258. 14 (2021) 1258. https://doi.org/10.3390/MA14051258.
82. R.S. Figliola, D.E. Beasley, Theory and design for mechanical measurements, (n.d.).rns, Opt. Laser Technol. **30**, 27–31 (1998). https://doi.org/10.1016/S0030-3992(97)00062-5
83. M. Rabiei. Development and analysis of nanocomposite based on hydroaxypatite for application in bioengineering. Doctoral thesis, Kaunas University of Technology, (2023)

Chapter 2
Assessing Techniques for Determining Nano Crystal Size of Natural Hydroxyapatite via X-Ray Diffraction

This chapter discusses a comparative study of various methods to determine crystal size from XRD patterns [1]. It focuses on XRD peaks extracted from hydroxyapatite derived from bones of cows, pigs, and chickens. The synthesis of hydroxyapatite involved thermally processing natural bones at 950 °C. X-Pert software was employed to tailor XRD patterns for each method and compute crystal sizes. The methods examined include Scherrer (three models), Monshi–Scherrer, three Williamson–Hall models (the Uniform Deformation Model (UDM), the Uniform Stress Deformation Model (USDM), and the Uniform Deformation Energy Density Model (UDEDM)), Halder–Wagner (H-W), and the Size Strain Plot Method (SSP). These approaches were evaluated and compared. Crystal sizes determined from XRD patterns for hydroxyapatite from cow, pig, and chicken were 1371, 457, and 196 nm using the Scherrer method (straight line model across all peaks). A revised model (straight line through the origin) yielded sizes of 60, 60, and 53 nm, indicating notable improvement. Average sizes were found to be 56, 58, and 52 nm for each method's variant, respectively, for cow, pig, and chicken. The Monshi–Scherrer method returned sizes of 60, 60, and 57 nm. The UDM method yielded sizes of 56, 62, and 65 nm. USDM calculated sizes were 60, 62, and 62 nm. UDEDM produced sizes of 62, 62, and 65 nm for cow, pig, and chicken, respectively. In the H-W method, crystal sizes were consistently 4 nm across all samples. The SSP method estimated sizes at 43, 62, and 57 nm for cow, pig, and chicken, respectively. Upon comparing these methods, the Scherrer method (straight line model) across all peaks yielded implausible values; however, other methods produced sizes within acceptable ranges as corroborated by existing literature. Experimental analyses, such as specific surface area via gas adsorption (Brunauer–Emmett–Teller (BET)) and Transmission Electron Microscopy (TEM), were conducted. The final comparison focused on accuracy, simplicity of calculation, researcher checkpoints, and the variance between calculated and experimental BET and TEM values. The Monshi–Scherrer method facilitated easier calculations and reduced error through least squares applied to linear plots, ensuring the slope is close to one and deriving the intercept for precise crystal sizes.

M. Rabiei et al., *Hydroxyapatite-Based Nanocomposites*, Springer Series in Chemical Physics 127, https://doi.org/10.1007/978-3-032-12583-5_2

This study found the BET values (56, 52, and 49 nm) closely matched those from the Monshi–Scherrer method, recommending its use in nanotechnology research.

2.1 Introduction

A crystallite solid is characterized by a periodic assembly of atoms, molecules, or ions. In crystalline solids, the terms "order" and "disorder" refer to variations in this periodicity and arrangement [2]. X-ray diffraction (XRD) profile analysis is an effective means of examining crystallite size and lattice strain. Calculating nanocrystallite size, particularly within nanocrystalline bulk materials, is facilitated by X-ray diffraction patterns and crystallographic techniques. Paul Scherrer's 1918 paper introduced the Scherrer equation [3], which Uwe Holzwarth and Neil Gibson later associated with sharp X-ray diffraction peaks. This equation, specifically related to single (hkl) peaks, is applicable only for average sizes up to approximately 100 nm and is influenced by instrument specifications and signal-to-noise ratios. As crystallite size increases, the breadth of diffraction peaks decreases, complicating the differentiation between broadening due to crystallite size and other factors. Optimal calculation methods mitigate errors to enhance accuracy. Given that nanoparticles comprise multiple nano- or microscale crystals, size determination from XRD patterns pertains to crystals rather than particles. This study analyzes XRD peak lists from hydroxyapatite Hydroxyapatite samples derived from cow, pig, and chicken bones to evaluate various methods for crystal size estimation via XRD. Seven XRD peak-related methods are applied, and a novel Scherrer method model (a straight line intersecting the origin) is proposed. Interest in nanomaterials from biological sources like bovine hydroxyapatite is rising, particularly for applications in tissue engineering and implant fabrication, due to their distinct properties when used in conjunction with metals, bioglasses, polymers, and composites.

2.2 Materials and Experiments

Bones from cows, pigs, and chickens obtained from a Maxima LT shop according to EU regulations on Lithuanian breeds were initially boiled in hot water for two hours to remove surface meats and fats. Following this, they were cleaned and dried at 110 °C for two hours. Subsequent thermal treatment of hydroxyapatite Hydroxyapatite in a furnace set at 950 °C for two hours facilitated the diffusion of proteins, like collagens, from the bone interiors to the surface and involved combustion at elevated temperatures. The furnace used was an E5CK-AA1-302 model (Snol 6, 7/1300). In this examination, a D8 Discover X-ray diffractometer (Bruker AXS GmbH, Kaunas, Lithuania) with $Cu_{K\alpha}$ radiation was employed. This process yielded a white and purified hydroxyapatite. Samples were subsequently ground in a rotary ball mill with a specified ratio of fired grogs, steel balls, and void space. The ball mill utilized was a planetary Fritsch Pulverisette-5 (Kaunas, Lithuania). The particles were micron-sized, while the crystals within the particles were nanoscale, as documented in the literature [5]. Powder X-ray diffractions were captured at 40 kV and 40 mA, recorded over a range of 20 to 50 degrees for 2θ at a scanning rate of 2.5 degrees/minute with a step increment of 0.02 degrees. The resultant diffraction patterns were analyzed using version 4.9 of the High Score X'Pert software, employing the fundamental parameter method integrated into ASC suffix files. Furthermore, the specific surface area of the samples was assessed using the desorption isotherms of nitrogen (N_2) gas through a Brunauer–Emmett–Teller (BET) apparatus, specifically the Gemini V analyzer, micrometrics GmbH (Isfahan, Iran). For elemental analysis, an energy dispersive X-ray (EDX) spectrometer, Philips/FEI Quanta 200, was utilized. Additionally, for examining thin sample layers, transmission electron microscopy (TEM) using the CM 10-Philips (Tehran, Iran) operating at acceleration voltages between 50 and 80 kV was used.

2.2.1 Preparation of Hydroxyapatite Powders

The femurs of cows, pigs, and chickens were readied for processing. Initially, bovine bones were separated and boiled in hot water, followed by a two-hour soak in acetone to eliminate collagen and fat (step 1, Fig. 2.1). Subsequently, in step 2, these bones were rinsed with distilled water and dried twice. They were then incrementally heated in a furnace under ambient conditions at a rate of 10 °C per minute. Ultimately, the bones were heated at 950 °C for 2 h, then allowed to cool gradually within the furnace. This process transformed the initially black bones (due to carbon release) into a white granular material. Furthermore, at 950 °C, the bones fully crystallized into hydroxyapatite Hydroxyapatite (step 3) [6]. Following heat treatment, the hydroxyapatite derived from cow, pig, and chicken bones was processed in a planetary ball mill with a tungsten carbide bowl and balls to produce fine particles. The milling process involved a feed ratio of 30 g of powder to 300 g of balls (a weight ratio of 1:10), operated at a fixed speed of 250 rpm, with a milling duration of 2 h including pause and reverse modes (step 4), as outlined in previous studies [7, 8]. Production route images of hydroxyapatite are shown in Fig. 2.1. The phase composition and purity were analyzed by X-ray diffraction (XRD), with the resulting XRD patterns of the milled hydroxyapatite white powder depicted in Fig. 2.1. These patterns underwent thorough analysis via the X'Pert software and were matched with the standard XRD peaks of hydroxyapatite per ICDD 9–432. Similar findings have been recorded by Bahrololoom and Shahabi [9, 10]. Additionally, the crystallographic parameters for each XRD pattern can be found in Tables 2.1, 2.2, 2.3. Crystallographic parameters derived from X'Pert software are listed in Table 2.4. The unit cell parameters showed strong agreement with existing studies on hydroxyapatite fabrication [11, 12].

X-ray diffraction was employed to ascertain both the phase composition and the purity of the materials. The XRD patterns for the ground white HA powder are depicted in Fig. 2.3. These patterns underwent comprehensive analysis with the

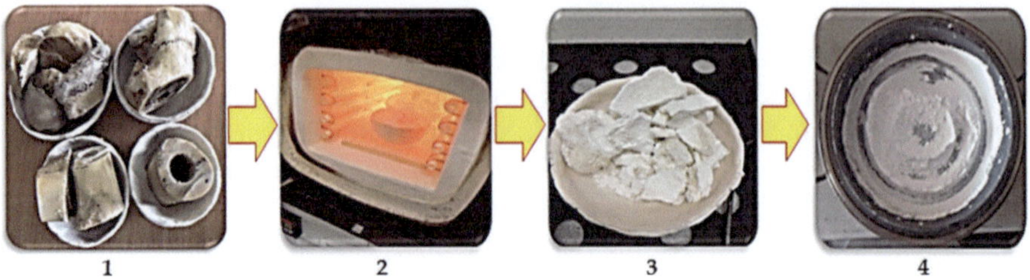

Fig. 2.1 Pictures of the production process of HA from bovine, porcine and chicken bones (steps 1–4)

X'Pert software and were validated against the standard XRD peaks for HA as per ICDD 9–432. Additionally, the crystallographic parameters associated with each XRD pattern are presented in Table 1, while the structures derived from the X'Pert software analysis are detailed in Table 2.1. The unit cell parameters correspond well with those reported by other researchers in HA fabrication.

Analysis of the XRD patterns (Fig. 2.2) revealed that the crystallization behavior of HA samples was largely consistent. The patterns were analyzed at angles spanning $20° < 2\theta < 50°$. The most prominent peaks linked to crystalline HA were detected at approximately 31.96°, 32.04°, and 32.03° for samples derived from bovine, porcine, and chicken sources, respectively. These pronounced peaks at specific 2θ values are indicative of the HA structure, with corresponding hkl values of 002, 102, 210, 211, 112, 300, and 202. Furthermore, the full width at half maximum (FWHM) values of these peaks, measured in radians, ranged from 0.00174 to 0.00348, 0.00226 to 0.00313, and 0.00244 to 0.00313 for HA extracted from cow, pig, and chicken bones, respectively (as shown in Table 2). Additionally, the peak intensities across samples were similar, holding steady around ~250 counts, which can be attributed to the comparable origin and type of the HA samples.

2.3 Results and Discussions

2.3.1 Scherrer Method

Equation (2.1) describes the Scherrer equation, linking it to the diffraction peak, as stated in reference [3]. In this equation, L represents the size of the nanocrystal, and K denotes the shape factor, typically 0.89 for ceramic substances. The wavelength of the radiation, denoted as λ, is measured in nanometers (with $\lambda(Cu_{K\alpha})$ being 0.15405 nm). The angle of diffraction is given by Θ, while β indicates the full width at half maximum of the peak, expressed in radians. Furthermore, the broadening of peaks can be attributed to both physical and instrumental causes, as per references [14, 15].

Table. 2.1 Crystallographic parameters of the XRD pattern related to the hydroxyapatite obtained from cow

Cow											
2θ (Degree)	β = FWHM (Degree)	θ (Degree)	cos θ (Degree)	l/cos θ (Degree)	Ln(l/cos θ) (Degree)	β = FWHM (Radian)	Lnβ (Radian)	4 sin θ (Degree)	B(Radian).cosθ (Degree)	hkl	dhki(Å)
26.15	0.14	13.07	0.9740	1.02669	0.02634	0.00244	-6.0174	0.9045	0.00238	002	3.46500
28.32	0.2	14.16	0.9696	1.03135	0.03087	0.00348	-5.66072	0.9785	0.00337	102	3.17485
29.18	0.1	14.59	0.9677	1.03338	0.03283	0.00174	-6.35387	1.007	0.00168	210	3.07687
31.966	0.15	15.98	0.9613	1.04026	0.03947	0.00261	-5.94841	1.1012	0.00251	211	2.81215
32.54	0.14	16.27	0.9599	1.04178	0.04093	0.00244	-6.0174	1.1206	0.00234	112	2.78900
32.98	0.15	16.49	0.9588	1.04297	0.04207	0.00261	-5.94841	1.3687	0.00245	310	2.26285
33.97	0.14	16.98	0.9564	1.04559	0.04458	0.00244	-6.0174	1.1681	0.00233	202	2.63845
40.03	0.15	20.01	0.9396	1.06428	0.0623	0.00261	-5.94841	1.3687	0.00245	310	2.26285
46.94	0.16	23.47	0.9172	1.09027	0.08643	0.00278	-5.88387	1.5930	0.00255	222	1.94339
48.35	0.2	24.17	0.9123	1.09613	0.09179	0.00348	-5.66072	1.6377	0.00317	320	1.87176
49.73	0.15	24.86	0.9073	1.10217	0.09728	0.00261	-5.94841	1.6816	0.00237	213	1.84732

Table. 2.2 Crystallographic parameters of the XRD pattern related to the hydroxyapatite obtained from pig

Pig

2θ (Degree)	β = FWHM (Degree)	θ (Degree)	cos θ (Degree)	l/cos θ (Degree)	Ln(l/cos θ) (Degree)	β = FWHM (Radian)	Lnβ (Radian)	4 sin θ (Degree)	B(Radian).cosθ (Degree)	hkl	dhki(Å)
26.12	0.13	13.06	0.9741	1.02659	0.0224	0.00226	-6.09151	0.9038	0.0022	002	3.46500
29.20	0.14	14.60	0.9677	1.03338	0.03283	0.00244	-6.0174	1.0082	0.00236	210	3.07687
32.04	0.14	16.02	0.9611	1.04047	0.03968	0.00244	-6.0174	1.1038	0.00235	211	2.81215
32.44	0.13	16.22	0.9601	1.04156	0.04072	0.00226	-6.09151	1.1173	0.00217	112	2.78900
33.07	0.14	16.53	0.9586	1.04319	0.04228	0.00244	-6.0174	1.1380	0.00234	300	2.71354
34.02	0.14	17.01	0.9562	1.04581	0.04479	0.00244	-6.0174	1.1701	0.00233	202	2.63845
40.07	0.18	20.03	0.9395	1.0644	0.06241	0.00313	-5.76608	1.3700	0.00294	310	2.26285
46.96	0.15	23.48	0.9171	1.09039	0.08654	0.00261	-5.94841	1.5937	0.00239	222	1.94339
48.34	0.14	24.17	0.9123	1.09613	0.09179	0.00244	-6.0174	1.6377	0.00223	320	1.87176
49.73	0.15	24.86	0.9073	1.10217	0.09728	0.00261	-5.94841	1.6816	0.00237	213	1.84732

Table. 2.3 Crystallographic parameters of the XRD pattern related to the hydroxyapatite obtained from chicken

Chicken											
2θ (Degree)	β = FWHM (Degree)	θ (Degree)	cos θ (Degree)	l/cos θ (Degree)	Ln(l/cos θ) (Degree)	β = FWHM (Radian)	Lnβ (Radian)	4 sin θ (Degree)	B(Radian).cosθ (Degree)	hkl	dhki(Å)
26.20	0.14	13.10	0.9739	1.0268	0.02645	0.00244	-6.0174	0.9066	0.00238	002	3.46500
28.39	0.16	14.19	0.9694	1.03157	0.03108	0.00278	-5.88387	0.9805	0.00269	102	3.17485
29.19	0.15	14.59	0.9677	1.03338	0.03283	0.00261	-5.94841	1.0076	0.00253	210	3.07687
32.03	0.16	16.01	0.9612	1.04037	0.03957	0.00278	-5.88387	1.1032	0.00267	211	2.81215
32.45	0.15	16.22	0.9601	1.04156	0.04072	0.00261	-5.94841	1.1414	0.0025	300	2.71354
34.21	0.16	17.10	0.9557	1.04635	0.04531	0.00278	-5.88387	1.1761	0.00266	202	2.63845
40.05	0.17	20.02	0.9395	1.0644	0.06241	0.00296	-5.82324	1.3693	0.00278	310	2.26285
46.95	0.18	23.47	0.9172	1.09027	0.08643	0.00313	-5.76608	1.5930	0.00287	222	1.94339
48.34	0.17	24.17	0.9123	1.09613	0.09179	0.00296	-5.82324	1.6377	0.0027	320	1.87176
49.74	0.18	24.87	0.9072	1.10229	0.09739	0.00313	-5.76608	1.6822	0.00284	213	1.84732

Table. 2.4 Crystallographic parameters related to the hydroxyapatite structure resulting via X'Pert software

Bone	Crystal System	a (Å)	c (Å)	c/a (Å)	Cell Volume ($Å^3$)	Crystal Density (g/cm^3)
Cow	Hexagonal	9.4000	6.9300	0.7340	530.30	3.14
pis	Hexagonal	9.4210	6.8930	0.7316	529.83	3.14
Chicken	Hexagonal	9.4210	6.8800	0.7302	528.83	3.18

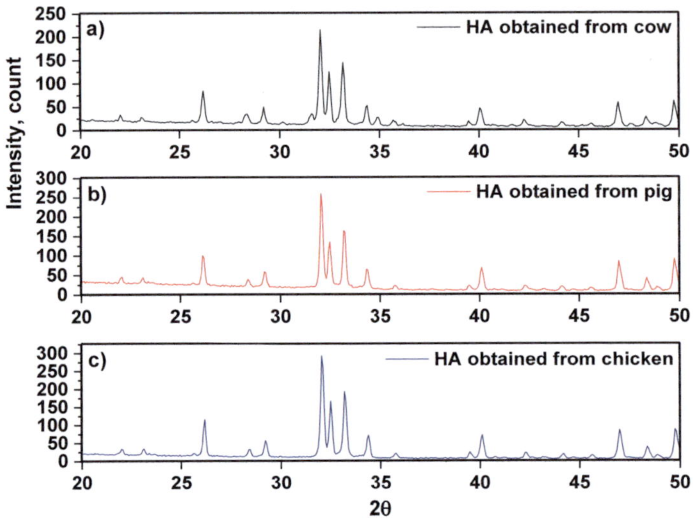

Fig. 2.2 XRD patterns of HA obtained from **a** cow, **b** pig, and **c** chicken bones

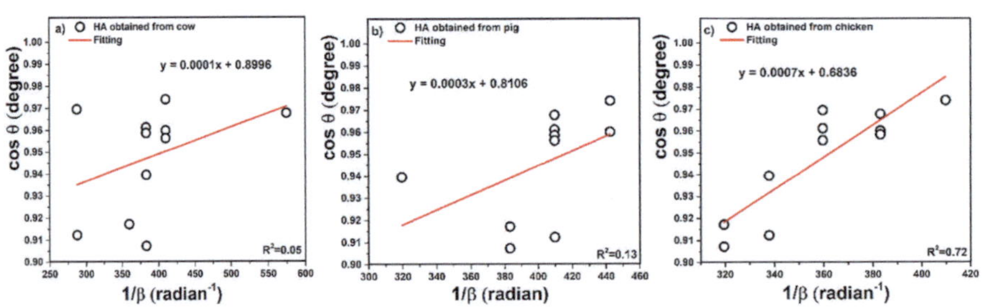

Fig. 2.3 Pattern of XRD analysis of hydroxyapatite obtained from **a** cow, **b** pig, and **c** chicken bones

$$L = \frac{K\lambda}{\beta} \cdot \frac{1}{\cos\theta} \tag{2.1}$$

For decreasing this error of instrument, Eq. (2.2) can be used:

$$\beta_d^2 = \beta_m^2 - \beta_i^2 \tag{2.2}$$

In this formula, βm is the measured broadening, βi is the instrumental broadening, and βd was introduced as the corrected broadening responsible for crystal size. Furthermore, in this case, crystalline silicon was used as the reference material for calibration of instrumental error. The sample's instrumental broadening and physical broadening are assessed via the full width at half maximum (FWHM). By applying a correction for physical broadening, the Scherrer equation can subsequently be used to calculate crystal size, as referenced in [16, 17]. Some publications have only applied the Scherrer equation to the sharpest peak without evaluating all or specific peaks.

2.3.1.1 Straight Line Model in Scherrer Method

In this instance, every peak was taken into account, and in alignment with the Scherrer equation, Fig. 2.3 illustrates plots of cosθ against 1/β (measured in inverse radians) for the respective samples. This approach utilizes a linear model, enabling the concurrent use of either all or selected peaks.

According to Eq. (2.3), the slope of plots is equal to Kλ/(L), therefore, the values of slope reported as 0.0001 for cow, 0.0003 for pig, and 0.0007 for chicken tandemly, after the calculation, gave values of crystal size as 1371, 457, and 196 nm for cow, pig, and chicken, respectively.

$$\cos\theta = \frac{K\lambda}{L} \cdot \frac{1}{\beta} \tag{2.3}$$

It is clear that since hydroxyapatite in natural bone exists as nanocrystals, crystallite sizes derived from the linear fit slope are incorrect, given that they must be smaller than 100 nm. When employing the least squares method to fit data to the Scherrer equation (Eq. (2.3)), the y-intercept lacks any physical significance. To improve the accuracy of the Scherrer equation application, it is advised to constrain the linear plot to intersect the origin.

2.3.1.2 Model of Straight Line Passing the Origin in Scherrer Method

This study presents a newly developed model. To ensure that the linear plot intersects the origin and provides an accurate slope for computational purposes, Eq. (2.4) [18] was utilized. All data points were extracted as a plot of y against x, depicted in Fig. 2.3, and are detailed in Table 2.5.

$$\text{Slope} = \frac{x_1y_1 + x_2y_2 + x_3y_3 + \cdots \ldots\ldots + x_ny_n}{x_1^2 + x_2^2 + x_3^2 + \ldots\ldots\ldots\ldots\ldots + x_n^2} \tag{2.4}$$

Table. 2.5 The (x,y) points extracted by the plots in Fig. 2.3

Cow		Pig		Chicken	
X	y	X	y	X	y
409.83	0.974	442.47	0.9741	409.83	0.9739
287.35	0.9696	409.83	0.9677	359.71	0.9694
574.71	0.9677	409.83	0.9611	383.14	0.9677
383.14	0.9613	442.47	0.9601	359.71	0.9612
409.83	0.9599	409.83	0.9586	383.14	0.9601
383.14	0.9588	409.83	0.9562	383.14	0.9584
409.83	0.9564	319.48	0.9395	359.71	0.9557
383.14	0.9396	383.14	0.9171	337.83	0.9395
359.71	0.9172	409.83	0.9123	319.48	0.9172
287.35	0.9123	383.14	0.9073	337.83	0.9123
383.14	0.9073	-	-	319.48	0.9072

Following the computations, the resulting slope measurements were 0.0023, 0.0023, and 0.0026 for the various samples. Consequently, the crystal sizes determined were 60 nm, 60 nm, and 53 nm for cow, pig, and chicken, respectively. This adaptation, made in this study, applies the Scherrer equation concurrently to all peaks.

2.3.1.3 Average Model in Scherrer Equation

A typical model based on the Scherrer equation was employed; the crystal size was computed using Eq. (2.1) and subsequently averaged. Crystal size values derived using the average method with the Scherrer equation are detailed in Table 2.6. The average sizes for cow, pig, and chicken are 56, 58, and 52 nm, respectively. Hence, it appears that the crystallite size estimated from the slope by modifying the linear fit, as done in this study (case 2), shows greater consistency compared to using the Scherrer equation to individual peaks and then averaging. This could be attributed to the observation that as the diffraction angle rises from 25 to 50 (Fig. 2.2), the FWHM values become less reliable [19], whereas averaging presumes equal validity for all points. Employing Eq. (2.4) [18] uses a least squares method for a linear graph that must intersect the origin, thus adjustments are made to minimize error sources.

2.3.2 *Modified Scherrer Equation (Monshi–Scherrer Method)*

Monshi et al. in 2012 employed some modifications in use of the Scherrer equation and introduced the following formula (Eq. (2.5)) [19]:

Table. 2.6 Values of crystal size extracted by the average method based on the Scherrer equation

$\frac{K\lambda}{\beta \cos\theta}$ of Cow	$\frac{K\lambda}{\beta \cos\theta}$ of Pig	$\frac{K\lambda}{\beta \cos\theta}$ of Chicken
57.60	62.32	57.60
40.68	58.09	50.96
81.60	58.34	54.19
54.62	63.18	51.35
58.59	58.59	54.62
54.84	58.84	54.84
58.84	46.63	51.54
55.96	57.36	49.31
53.76	61.4S	47.77
43.25	57.85	50.77
57.85	-	48.27
56	58	52

The Scherrer equation systematically shows increased values of nanocrystalline size as d (distance of diffracted planes) values decrease and 2θ values increase, since β.cosθ cannot be maintained as constant. Furthermore, the Modified Scherrer equation can provide the advantage of decreasing the errors or $\Sigma(\pm\Delta\ln\beta)^2$ to give a more accurate value of L from all or some of the different peaks [19].

$$\mathrm{Ln}\beta = \mathrm{Ln}\left(\frac{K\lambda}{L}\right) + \mathrm{Ln}\left(\frac{1}{\mathrm{Cos}\theta}\right) \tag{2.5}$$

So that the linear plot of Ln β (β in radians) versus Ln $\left(\frac{1}{\mathrm{Cos}\,\theta}\right)$ (degree) can be a linear plot for all or some of the chosen peaks, the least squares statistical method is used to decrease the sources of errors. After stablishing the most accurate linear plot, the value of Ln $\left(\frac{K\lambda}{L}\right)$ can be obtained from the intercept. The $e^{(\text{intercept})}$ gives $\frac{K\lambda}{L}$, from which a single value of L is obtained from all of the available peaks. The plots in Fig. 4 illustrate the relationship between Lnβ and ln(1/cosΘ), alongside the linear least squares equations derived from data using linear regression. To determine crystal sizes, the Monshi–Scherrer equation, denoted as Eq. (6), is applied. For optimal use with X'Pert software, it's recommended to generate an ASC file of peak data (with the suffix ASC) to create a peak list that includes FWHM, accessible via the fit profile option (right-click the peak and select fit profile in X'Pert). This enables comprehensive fitting to determine β (FWHM). By plotting Eq. (2.5) and developing a linear equation based on the least squares method for either all peaks or a chosen selection,

$$\frac{K\lambda}{L} = e^{(\text{intercept})} \tag{2.6}$$

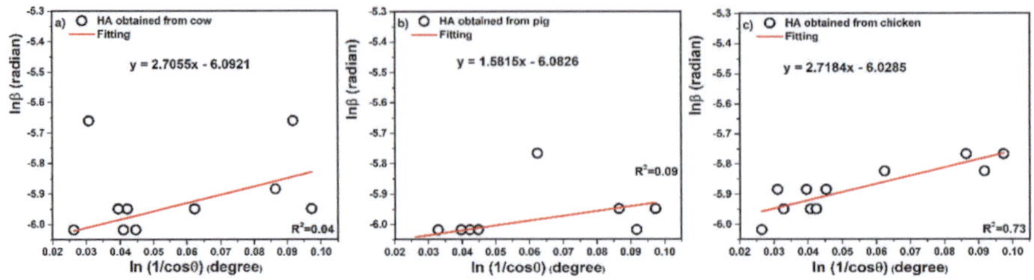

Fig. 2.4 Linear plots of the modified Scherrer (Monshi–Scherrer) equation and gained intercepts for different hydroxyapatites obtained from (**a**) cow, (**b**) pig, and (**c**) chicken bones

Linear equations of hydroxyapatite obtained from cow, pig, and chicken recorded $y = 2.7055x - 6.0921$, $y = 1.5815x - 6.0826$, and $y = 2.7184x - 6.0285$, respectively, and intercept values were -6.0921 for cow, -6.0826 for pig, and -6.0285 for chicken tandemly. Nevertheless, the intercepts were calculated as $e^{(-6.0921)} = 0.00227$, $e^{(-6.0826)} = 0.00228$, and $e^{(-6.0285)} = 0.00240$, respectively. Therefore, $\frac{K\lambda}{L} = 0.00227$, $\frac{K\lambda}{L} = 0.00228$ and $\frac{K\lambda}{L} = 0.00240$ for cow, pig, and chicken tandemly. After the calculations, the values of crystal sizes were obtained as 60, 60, and 57 nm for cow, pig, and chicken, respectively Fig. 2.4.

2.3.3 *Williamson–Hall Method of Analysis*

The Scherrer equation primarily accounts for the influence of crystallite size on the broadening of XRD peaks; however, it does not account for the microstructural aspects of the lattice, specifically the intrinsic strain, which can emerge in nanocrystals due to point defects, grain boundaries, triple junctions, and stacking faults [20]. Alternatively, the Williamson–Hall (W–H) method considers the influence of strain-related broadening of XRD peaks and allows for the computation of both crystal size and intrinsic strain [21, 22]. X-ray diffraction peak broadening fundamentally results from a combination of size and strain effects. Unlike the Scherrer equation, the W–H method aligns with a dependency on tanθ, rather than 1/cosθ, when accounting for strain. This fundamental distinction separates the contributions of size and strain to peak broadening, effectively integrating microstrains with small crystallite sizes. The specific θ relational aspects involved in examining size and strain broadening in W–H are represented by Eq. (2.7).

$$\beta_{total} = \beta_{size} + \beta_{strain} \tag{2.7}$$

In this case, modified W–H was used and models involved uniform deformation (UDM), uniform stress deformation (USDM), and uniform deformation energy density (UDEDM), which will be discussed.

2.3.3.1 Uniform Deformation Model (UDM)

The UDM method obtained the following equation (Eq. (2.8)) for the strain associated with the nanocrystals:

$$\varepsilon = \frac{\beta}{4\tan\theta} = \frac{\beta_2 \cos\theta}{4\sin\theta} \tag{2.8}$$

where β_2 is the broadening of the width of the peaks due to strain, while the broadening due to nanocrystal size β_1 comes from the Scherrer equation.

$$\beta = \beta_1 + \beta_2 = \frac{K\lambda}{L\cos\theta} + 4\varepsilon\frac{\sin\theta}{\cos\theta} \tag{2.9}$$

$$\beta_{hkl} \cdot \cos\theta = \left(\frac{K\lambda}{L}\right) + (4\varepsilon\sin\theta) \tag{2.10}$$

According to Eq. (2.10), the term of ($\beta_{hkl}cos\theta$) corresponds to (4 $sin\theta$) for the preferred orientation peaks of hydroxyapatite with the hexagonal lattice and considers the isotropic nature of the crystals. Figure 2.5 shows the 4 sinΘ as an X-axis and β cosθ term along the Y-axis. Mostly, UDM is related to an isotropic (perfect) crystal system in all (hkl) planes. Apparently, slope and intercept of the fitted line correspond to the strain and crystal size, respectively. The intercept values equal $\frac{K\lambda}{L}$. The $\frac{K\lambda}{L}$ reported was 0.0021 for cow, 0.0022 for pig, and 0.0021 for chicken. These quantities are estimated from the intercept of the vertical axis and slope, from the plot of $\beta_{hkl}cos\theta$ as a function of 4 $sin\theta$. After calculations, the crystal size values were obtained as 65, 62, and 65 nm for cow, pig, and chicken, respectively. In this plot, the units of 4 sinΘ and β cosθ are degree and radian degree tandemly. In addition, several defects influence to the lattice structure via size restriction and it will be caused to the strain lattice.. Herein, the slope values (positive values) are represented to the intrinsic strain, therefore, 0.0003 for cow, 0.0002 for pig, and 0.0004 for chicken have been reported. The positive values of intrinsic strain can prove tensile strain and if values were negative, they will be related to the compressive strain.

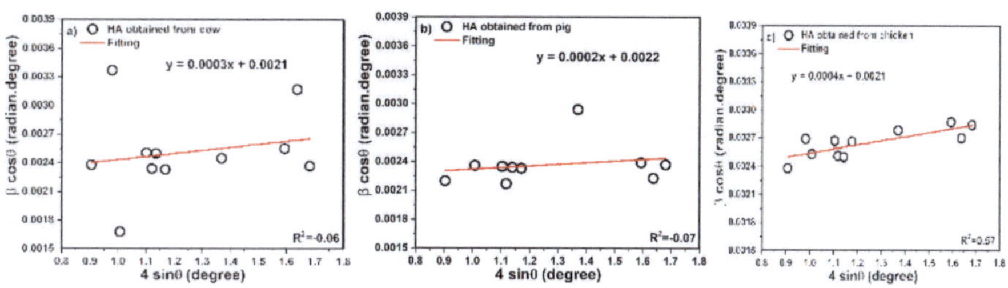

Fig. 2.5 UDM plot of hydroxyapatite obtained from **a** cow, **b** pig, and **c** chicken bones

2.3.3.2 Uniform Stress Deformation Model (USDM)

In a more accurate crystal model that accounts for the anisotropic properties of Young's modulus [23], Hooke's law is extended to describe the linear relationship between strain (ε) and stress (σ), with Young's modulus serving as the proportionality constant. This technique involves applying Hooke's law to relate the crystal's strain and stress linearly, as outlined in Eq. (2.11), where σ represents stress, ε denotes strain, and E symbolizes Young's modulus, respectively.

$$\sigma = \mathrm{E}\varepsilon \tag{2.11}$$

This equation is applicable primarily to scenarios involving very small strains, where it holds significant credibility. Hooke's law can be applied when imaging these minimal strains. However, as strain increases, particles begin to diverge from linear behavior, as noted in [24]. To determine the modulus E, one can refer to Eq. (2.12), which is specific to the lattice type. For instance, as per data obtained from X'Pert, hydroxyapatite exhibits hexagonal crystal structures, mandating the use of Eq. (2.12) [25]. In this equation, the indices h, k, and l denote the crystallographic plane, while a and c represent lattice parameters, which can be retrieved from a phase file using X-pert software. In addition, S_{11} S_{33}, S_{13}, and S_{44} are introduced as elastic compliances and C_{11}, C_{12}, C_{33}, and C_{44} are elastic stiffness constants of hexagonal hydroxyapatite. The values of S_{11}, S_{33}, S_{13}, and S_{44} for hydroxyapatite are presented in Table 2.7 and the values are cited in reference [26]. Moreover, Table 2.8 displays the crystallographic parameters and Young's modulus (E) for each specific XRD pattern corresponding to hydroxyapatite derived from cow, pig, and chicken sources. Notably, the Young's modulus (Ehkl) is aligned perpendicular to the series of crystal lattice planes denoted as (hkl).

$$\mathrm{E_{hkl}} = \frac{\left[\mathrm{h}^2 + \frac{(\mathrm{h}+2\mathrm{k})^2}{3} + \left(\frac{\mathrm{al}}{\mathrm{c}}\right)^2\right]^2}{\mathrm{S_{11}}\left(\mathrm{h}^2 + \frac{(\mathrm{h}+2\mathrm{k})^2}{3}\right)^2 + \mathrm{S_{33}}\left(\frac{\mathrm{al}}{\mathrm{c}}\right)^4 + (2\mathrm{S_{13}} + \mathrm{S_{44}})\left(\mathrm{h}^2 + \frac{(\mathrm{h}+2\mathrm{k})^2}{3}\right)\left(\frac{\mathrm{al}}{\mathrm{c}}\right)^2} \tag{2.12}$$

According to Eq. (2.13), the terms of $\frac{4\mathrm{Sin}\theta}{\mathrm{E_{hkl}}}$ along the *X*-axis and $\beta_{hkl}.\cos\theta$ along the *Y*-axis are related to the peaks in the XRD pattern of the samples and are presented in Fig. 6.

Table. 2.7 Elastic compliances and stiffness constants of hydroxyapatite [26]

Elastic Compliances (GPa)					Stiffness Constants (GPa)				
C_{11}	**C_{12}**	**C_{13}**	**C_{33}**	**C_{44}**	**S_{11}**	**S_{12}**	**S_{13}**	**S_{33}**	**S_{44}**
137	42.5	54.9	172	39.6	0.88	−0.18	−0.22	0.72	2.52

Table. 2.8 Young's modulus (E) of each individual XRD pattern related to the hydroxyapatite obtained from cow, pig, and chicken bones

Cow		Pig		Chicken	
2Θ (Degree)	E (GPa)	2Θ (Degree)	E (GPa)	2Θ (Degree)	E (GPa)
26.15	138.889	26.12	138.889	26.20	138.889
28.32	123.935	29.20	113.636	28.39	124.121
29.18	113.636	32.04	108.694	29.19	113.636
31.96	108.734	32.44	113.02	32.03	108.684
32.54	112.887	33.07	113.636	32.45	113.054
32.98	113.636	34.02	110.706	33.16	113.636
33.97	110.598	40.07	113.636	34.21	110.733
40.03	113.636	46.96	107.155	40.05	113.636
46.94	107.161	48.34	113.636	46.95	107.154
48.35	113.636	49.73	112.702	48.34	113.636
49.73	112.571	-	-	49.74	112.734

$$\beta_{\mathrm{hkl}} \cdot \cos\theta = \left(\frac{\mathrm{K}\lambda}{\mathrm{L}}\right) + 4\sigma \cdot \frac{\sin\theta}{\mathrm{E_{hkl}}} \tag{2.13}$$

The size of crystals has been specified from the USDM method. The intercept values are equal to $\frac{K\lambda}{L}$, therefore, $\frac{K\lambda}{L}$ was reported as 0.0023 for cow, 0.0022 for pig, and 0.0022 for chicken. After calculation, the crystal size values were obtained as 60, 62, and 62 nm for cow, pig, and chicken, respectively. In addition, the slope of the straight line can provide values of stress, nevertheless, the values of stress for cow, pig, and chicken were calculated as 22, 18, and 44 MPa. The state of calculating strain was performed. In this step, average Young's modulus has further been reported. The average of the E value was calculated as 115.40 for cow, 114.58 for pig, and 115.45 GPa for chicken and values were not far from standard experimental value of Young's modulus (114 GPa([26]. Therefore, strain values were calculated as 1.89×10^{-4}, 1.59×10^{-4}, and 3.81×10^{-4} for cow, pig, and chicken, respectively Fig. 2.6.

2.3.3.3 Uniform Deformation Energy Density Model (UDEDM)

The study of anisotropic energy can be conducted through UDEDM. To determine the quantity of lattice energy conserved within a unit volume, the lattice energy density (LED) serves as a measuring piece. Here, we assume that the volumetric LED correlates with the effective stiffness of a crystal. According to Hooke's law, the LED can be derived from Eq. (2.14). Moreover, stress and strain are interconnected by Eq. (2.11), and the constants within the stress–strain relationship become interdependent when the strain energy density \(u \) is considered.

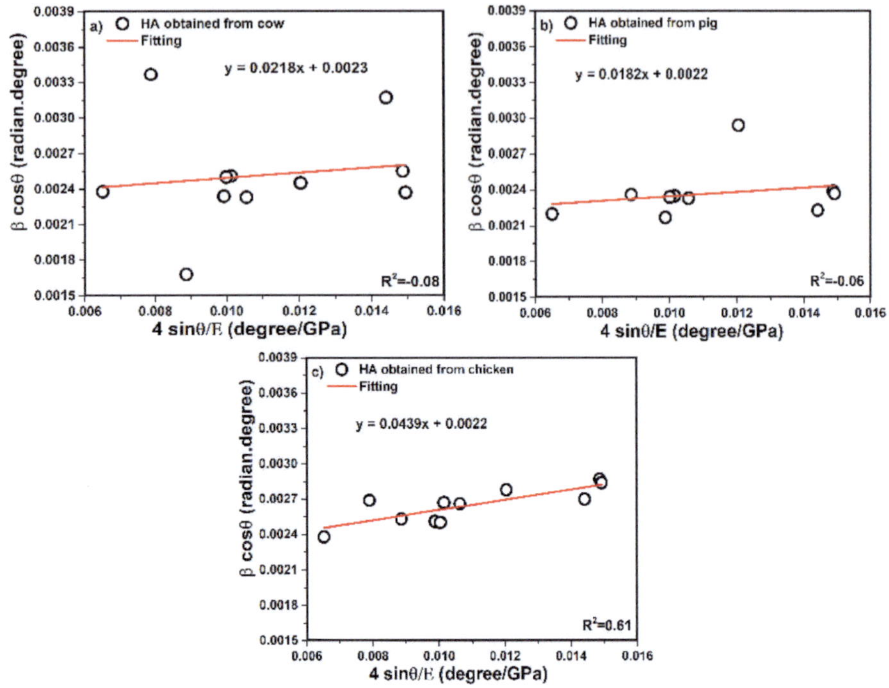

Fig. 2.6 USDM plot of hydroxyapatite obtained from (**a**) cow, **b** pig, and **c** chicken bones

$$\text{LED} = \varepsilon^2 \cdot \frac{\text{E}_{\text{hkl}}}{2} \tag{2.14}$$

Moreover, the intrinsic strain can be submitted as Eq. (2.15).

$$\varepsilon = \sigma \cdot \sqrt{\frac{2.\text{LED}}{\text{E}_{\text{hkl}}}} \tag{2.15}$$

Substitution of Eq. (2.15) with Eq. (2.10), yields Eq. (2.16) [27].

$$\beta_{\text{hkl}} \cdot \cos\theta = \left(\frac{\text{K}\lambda}{\text{L}}\right) + 4\sigma \cdot \sin\theta\sqrt{\frac{2 \cdot \text{LED}}{\text{E}_{\text{hkl}}}} \tag{2.16}$$

According to Eq. (2.16), values of crystal size can be calculated, as in Fig. 2.7, $\beta_{\text{hkl}}.\ \cos\Theta$ as a *Y*-axis and the term of $\frac{4\sin\theta}{\sqrt{\frac{\text{E}_{\text{hkl}}}{2}}}$ as an *X*-axis.

The intercept values of the plotted straight line equal $\frac{\text{K}\lambda}{\text{L}}$, therefore, $\frac{\text{K}\lambda}{\text{L}}$ was reported as 0.0022 for cow, 0.0022 for pig, and 0.0021 for chicken. In the final calculations, the crystal size values were reported as 62, 62, and 65 nm for cow, pig, and chicken tandemly. As a rule, in the application, compounds are not always isotropic,

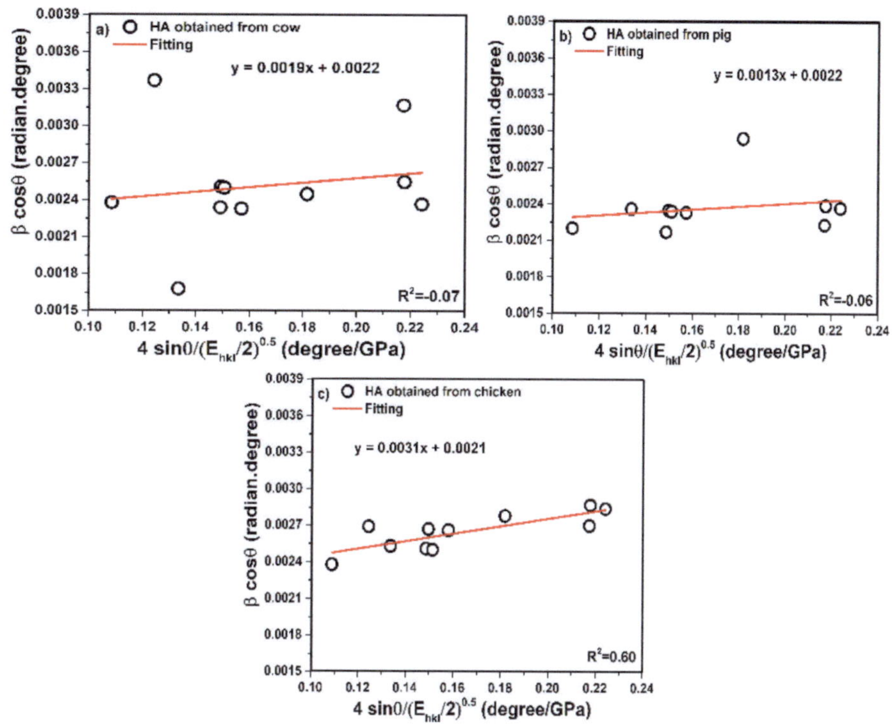

Fig. 2.7 UDEDM plot of hydroxyapatite obtained from **a** cow, **b** pig, and **c** chicken bones

perfect, and homogenous, but also compounds are encountered with defects, agglomeration, dislocations, imperfections, etc. In fact, another model obtained for energy density, where the constants of proportionality corresponded to strain–stress, is considered. In addition, two states similar to the USDM model were considered in calculations.

Moreover, the slope gives the LED, therefore, according to Eqs. (2.11) and (2.14), the values of energy density reported 43.67, 29.70, and 28.28 kJ/m^3 for cow, pig, and chicken tandemly. For strain values also, states have been noted. According to Eq. (2.14), the strain values were calculated as 0.87×10^{-3}, 0.73×10^{-3}, and 0.70×10^{-3} (E ~ average Young's modulus) for cow, pig, and chicken, respectively.

2.3.4 Halder–Wagner Method (H-W)

The core focus of this approach is based on the assumption that peak broadening follows a symmetric Voigt function [28, 29]. In line with this function, the complete width at half maximum of the physical profile is given by Eq. (2.17).

$$\beta_{hkl}^2 = \beta_L \cdot \beta_{hkl} + \beta_G^2 \tag{2.17}$$

Within this formula, βL and βG represent the full width at half maximum of the Lorentzian and Gaussian functions, respectively. A crucial point is the computation and the resulting values of the lattice spacing between the (hkl) planes, denoted as (d_{hkl}). For a hexagonal lattice structure like hydroxyapatite, this spacing is determined using Eq. (2.18). Conversely, for a cubic crystal lattice, the lattice spacing between the (hkl) planes corresponds to Eq. (2.19). The values of d_{hkl}) for hydroxyapatite derived from cow, pig, and chicken bones are documented in Tables 2.1–2.3.

$$\frac{1}{d_{hkl}^2} = \frac{4}{3}\left(\frac{h^2 + hk + k^2}{a^2}\right) + \left(\frac{l^2}{c^2}\right) \tag{2.18}$$

$$d_{hkl}^2 = \left(\frac{a^2}{h^2 + k^2 + l^2}\right) \tag{2.19}$$

Moreover, this approach concentrates on the peaks located at low to medium angles, where diffraction peak overlap is minimized. The Halder–Wagner calculation is detailed in Eq. (2.20), with its subformulas outlined in Eqs. (2.21) and (2.22) [30].

$$\left(\frac{\beta_{hkl}^*}{d_{hkl}^*}\right)^2 = \frac{1}{L} \cdot \frac{\beta_{hkl}^*}{d_{hkl}^*} + \left(\frac{\varepsilon}{2}\right)^2 \tag{2.20}$$

$$\beta_{hkl}^* = \beta_{hkl} \cdot \frac{\cos\theta}{\lambda} \tag{2.21}$$

$$d_{hkl}^* = 2d_{hkl} \cdot \frac{\sin\theta}{\lambda} \tag{2.22}$$

In addition, d_{hkl} is the lattice distance between the (hkl) planes for the hexagonal crystal, as well as the term of $\frac{\beta_{hkl}^*}{d_{hkl}^{*2}}$ for the X-axis and the term of $\left(\frac{\beta_{hkl}^*}{d_{hkl}^*}\right)^2$ for the Y-axis illustrated in Fig. 2.8.

The slope of the plotted line provides calculation of crystal size of samples. The slope values are proportional to $\frac{K\lambda}{L} = 0.0333$ for cow, $\frac{K\lambda}{L} = 0.0329$ for pig, and $\frac{K\lambda}{L} = 0.0322$ for chicken; after calculation, the crystal size values were obtained as 4, 4, and 4 nm for cow, pig, and chicken, respectively. In addition, calculated values of strain from the intercept of the plot equal $\left(\frac{\varepsilon}{2}\right)^2$, but according to the negative intercept, the following calculation of strain was not possible.

2.3.5 *Size Strain Plot Method (SSP)*

This approach assigns reduced importance to high-angle reflection data, which enhances outcomes for isotropic broadening. This is due to the fact that at elevated

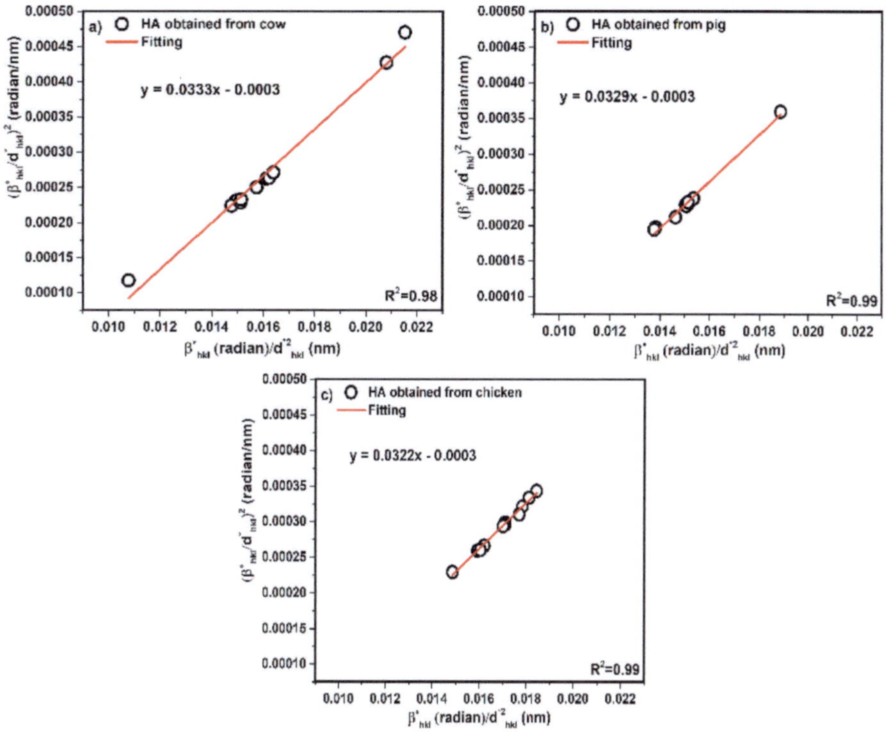

Fig. 2.8 Halder–Wagner plot of hydroxyapatite obtained from **a** cow, **b** pig, and **c** chicken bones

angles, diffracted XRD data tend to be of poorer quality with overlapping peaks. According to this model, the profile is depicted using a Gaussian function for the strain profile and a Lorentzian function for the crystallite size [31]. Moreover, the overall broadening associated with this method is represented by Eq. (2.23).

$$\beta_{hkl} = \beta_L + \beta_G \tag{2.23}$$

where, β_L and β_G are the peak broadening via Lorentz and Gaussian functions tandemly. Equation (2.24) is the submitted formula of the SSP method [27].

$$(d_{hkl} \cdot \beta_{hkl} \cdot \cos\theta)^2 = \frac{K\lambda}{L} \cdot \left(d_{hkl}^2 \cdot \beta_{hkl} \cdot \cos\theta\right) + \frac{\varepsilon^2}{4} \tag{2.24}$$

Figure 2.9 shows a plot of the $d_{hkl}^2 . \beta_{hkl} . \cos\theta$ term along the X-axis and $(d_{hkl} . \beta_{hkl} . \cos\theta)^2$ along the Y-axis corresponding to diffraction peaks.

The slope values are equal to $\frac{K\lambda}{L}$. The $\frac{K\lambda}{L}$ values are reported as 0.0032 for cow, 0.0022 for pig, and 0.0024 for chicken. After calculations, the crystal size values were obtained as 43, 62, and 57 nm for cow, pig, and chicken, respectively. According to Eq. (2.24), the intercept gives the intrinsic strain and the values of intercept are equal

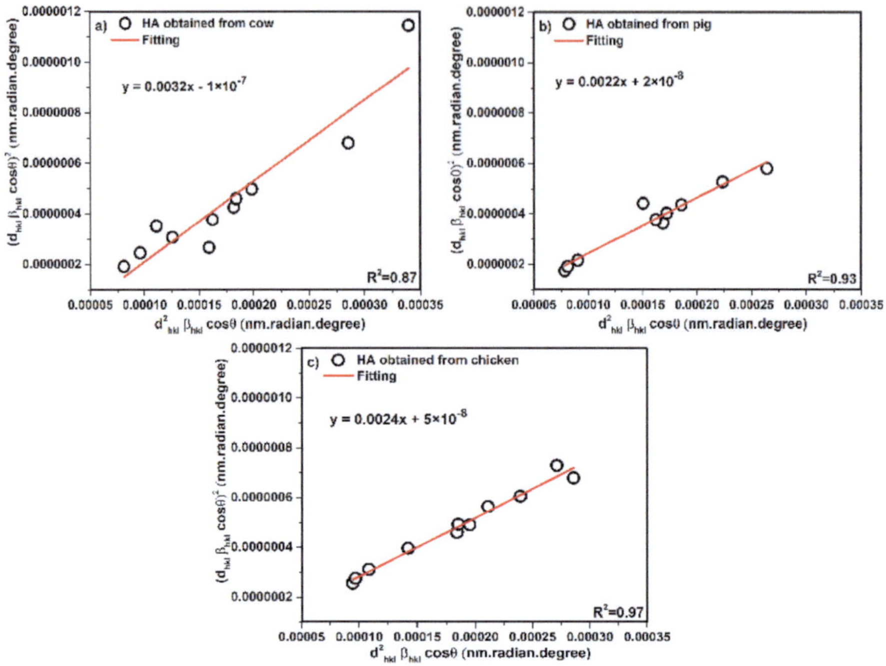

Fig. 2.9 SSP plot of hydroxyapatite obtained from **a** cow, **b** pig and **c** chicken bones

to $\frac{\varepsilon^2}{4}$, therefore, calculation of intrinsic strain values are possible for pig and chicken samples only because the intercept of cow is negative. The intrinsic strain values were calculated as 2.83×10^{-4} and 4.48×10^{-4} for pig and chicken, respectively.

2.3.6 *Specific Surface Area by Gas Adsorption (BET Method)*

The most widely used technique for estimating specific surface area is the BET method. Under normal atmospheric pressure and at the boiling temperature of liquid nitrogen, the amount of nitrogen adsorbed in relationship with pressure gives the specific surface area of powder. The observations are interpreted following the model of the BET method. The samples were degassed at 200 °C under reduced pressure (13×10^{-7} atmosphere) for around 15 to 20 h before each measurement. The amount of nitrogen was by volume adsorption at − 197 °C. The reported surface area for a bone-derived hydroxyapatite is much lower and the value is around 0.1 m^2/g [32]. However, one synthetic hydroxyapatite (not sintered or deproteinized bone) is 17 to 82 m^2/g [33]. Theoretical particle size can be calculated from adsorption specific surface area data by using Eq. (2.25).

$$D = \frac{6}{\rho \cdot S} \tag{2.25}$$

In this formula, ρ is the density of sample and S refers to the specific surface area of sample obtained from the BET method and similar to the method cited by Monshi et al. in [34]. The BET specific surface area of hydroxyapatite particles obtained from cow, pig, and chicken were 34.36 ± 0.01, 36.95 ± 0.01, and 43.39 ± 0.01 m^2/g. As explained, a theoretical particle size can be calculated from these data and the values of crystal size for hydroxyapatite calcined at 950 °C obtained from cow, pig, and chicken were 56, 52, and 49 nm, respectively.

2.3.7 *Study of TEM Analysis*

Figure 2.10 presents TEM images alongside the stoichiometric composition of hydroxyapatite nanocrystal powders sourced from cow, pig, and chicken bones following the ball milling procedure. According to the EDX analysis, the Ca/P ratios for hydroxyapatite derived from cow, pig, and chicken bones were determined to be 1.81, 1.79, and 1.68, respectively. A single particle can consist of multiple crystallites. Additionally, the TEM images reveal agglomerated nanocrystals, clearly depicting particle size, with crystals observable among all particles. Typically, the TEM size correlates with grain size; in this instance, it is evident that some powder particles have a nanoscale dimension, measuring less than 100 nm in width or diameter. A single particle, approximately 50 nm in size, is distinctly visible in chicken bone as shown in Fig. 2.10c. Moreover, the images display irregular spherical morphologies, which are referenced and verified in citation [36]. In addition, it can be related to the deviation of data points from line equations taking account to the fit (R^2), where there are differences between values of R^2 in each method. According to the calculation, R^2 allows it to be negative for some methods such as UDM (cow). It is intended to approximate the actual percentage variance, therefore, if the actual R^2 is close to zero, the R^2 can be slightly negative. However, the nanopowder was not dispersed perfectly, and the reason may be related to the agglomeration of powders created through the Van der Waals attraction [35]. In Fig. 2.10, there is spacing (d_{hkl}) with interplanar of less than 50 nm and the existence of hexagonal hydroxyapatite can be confirmed, according to the d_{hkl} reported in Tables 2.1–2.3. The results obtained from the methods and models are summarized in Tables 2.9 and 2.11. In addition, the crystal size of hydroxyapatite obtained from different natural sources in several studies are presented in Table 2.10, Table 2.11.

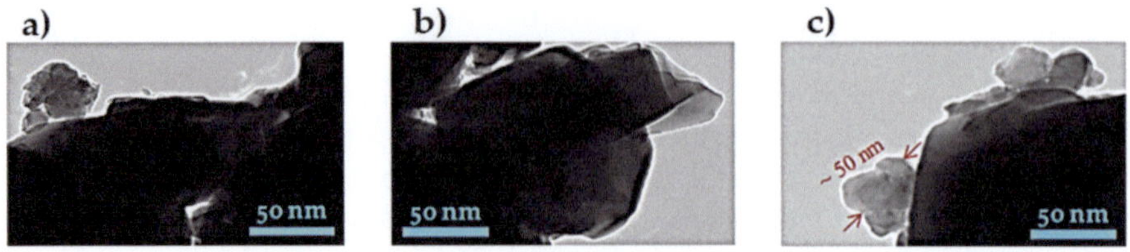

	O (Weight %)	P (Weight %)	Ca (Weight %)	Ca/P
HA obtained from cow bone	17.58	29.33	53.08	1.81
HA obtained from pig bone	19.32	28.94	51.73	1.79
HA obtained from chicken bone	16.87	31.01	52.11	1.68

Fig. 2.10 TEM images and stoichiometric composition of hydroxyapatite nanocrystals obtained from. **a** cow, **b** pig, and **c** chicken bones

Table. 2.9 Nanosize of hydroxyapatite crystallites obtained from cow, pig, and chicken bones extracted by some calculation methods and experimental methods (BET, TEM) in this study

Size of Crystallites	Scherrer (All Peaks/ New Model/ Average Model)	Monshi–Scherrer	Williamson–Hall (UDM/USDM/ UDEDM)	H-W	SSP	BET	TEM
L_{cow} (nm)	1371/60/ 56	60	65/60/62	4	43	56	~50
L_{pig} (nm)	457/60/ 58	60	62/62/62	4	62	52	~50
$L_{chicken}$ (nm)	196/53/ 52	57	65/62/65	4	57	49	~50

2.4 Conclusions

This research involved the successful synthesis of natural nano-hydroxyapatite derived from bovine, porcine, and avian bones. Various methods and models were compared for measuring the nanocrystallite size of these bones, alongside BET and TEM analyses. The Scherrer method data, when considering all diffraction peaks using a straightforward linear model, exceeded the nanoscale threshold of 100 nm. To address this, a new Scherrer model using a line passing through the origin is suggested, providing more accurate results when considering multiple or selective peaks simultaneously. Using an average model for each peak with the standard Scherrer formula produces reasonable sizes, but lacks a least squares approach to mitigate error sources. The Monshi–Scherrer method involves plotting Lnβ versus Ln(1/cosθ) and applying the least squares technique; the intercept provides Ln(Kλ/

Table. 2.10 Reports of some studies on the nanocrystallite size of hydroxyapatite prepared at high temperature

Number	Source	Method of Preparation	Temperature of Heat Treatment	Crystallite Phases	Size of Crystal (L) (nm)	Shape	Reference
1	Bovine bone	Thermal treatment	800 °C	hydroxyapatite	< 100 (58 and 62)	Needle	[36]
2	Bovine bone	Thermal treatment	800 °C	hydroxyapatite	70–180	Irregular	[37]
3	Fish scale	Thermal treatment	800 °C	hydroxyapatite	30	Irregular	[38]
4	Bovine bone	Thermal treatment	900 °C	hydroxyapatite	30	-	[39]
5	Bovine bone	Thermal treatment	900 °C	hydroxyapatite	70–80	Spherical	[40]
6	Pig bone	Thermal treatment	1000 °C	hydroxyapatite	38–52	Rod like	[41]
7	Fish scale	Thermal treatment	1000 °C	hydroxyapatite	76	Nearly spherical	[42]
8	Clam shell	Thermal treatment	1000 °C	hydroxyapatite	53–67	Agglomerate	[43]

Table. 2.11 Geometrical parameters of hydroxyapatite using different models in this study

	Williamson–Hall					SSP
	UDM	USDM		UDEDM		
	Strain (ε) × 10^{-4}	Stress (σ) (MPa)	Strain (ε) × 10^{-4}	Strain (ε) × 10^{-3}	LED (KJ/m^3)	Strain (ε) × 10^{-4}
Cow	3	22	1.89	0.87	43.67	–
Pig	2	18	1.59	0.73	29.70	2.83
Chicken	4	44	3.81	0.70	28.28	4.48

L), from which a single L value is derived. This method is straightforward, offering a checkpoint for researchers: the slope should be near one; otherwise, data from peaks at higher diffraction angles should be discarded for accuracy improvement. Crystal size data from the three Williamson–Hall method models are largely accurate and consistent, despite differing from BET experiment results which suggest smaller crystal sizes in chicken than in pig or cow bones. Additionally, strain and stress values display compressive behavior (positive values) across all (hkl) lattice planes. The variance in strain between UDM–USDM and UDEDM correlates with deformation uniformity. The Halder-Wagner method yielded crystal sizes of 4 nm for all samples, indicating significantly lower estimates. The lower accuracy of the H-W method can be related to the morphology of crystals, because $\frac{\beta^*_{hkl}}{d^{*2}_{hkl}}$ linearly increases with $\left(\frac{\beta^*_{hkl}}{d^*_{hkl}}\right)^2$ for all reflections with a positive slope and negative intercept, showing that there are no macrostrains and also morphology of crystals is not spherical crystallite shape, confirmed through the TEM images (irregular spherical). The report of strain by the H-W method was not possible because the intercept was negative and equal to $\left(\frac{\varepsilon}{2}\right)^2$. The SSP method yielded superior results compared to the H-W method, as H-W takes into account the influence of low to mid-angle XRD data and the effects of lattice dislocations. Ultimately, the Monshi–Scherrer method is recommended for both research and industrial purposes due to its simplicity, its role as a checkpoint for researchers, its use of the least squares method for enhanced accuracy, and its alignment with experimental BET data.

References

1. Rabiei, Marzieh; Palevicius, Arvydas; Monshi, Ahmad; Nasiri, Sohrab; Vilkauskas, Andrius; Janusas, Giedrius. Comparing methods for calculating nano crystal size of natural hydroxyapatite using X-Ray diffraction. Nanomaterials. Basel: MDPI. ISSN 2079–4991. 2020, **10**:9(127), p. 1–21. https://doi.org/10.3390/nano10091627
2. P.d.S. Santos, A.C.V. Coelho, H.d.S. Samohdntos, P.K. Kiyohara, Hydrothermal synthesis of well-crystallised boehmite crystals of various shapes. Mater. Res. **2009**, 12, 437–445, https://doi.org/10.1590/S1516-14392009000400012
3. Scherrer, P. Bestimmung der Grösse und der Inneren Struktur von Kolloidteilchen Mittels Röntgenstra-hlen. J. Nachrichten von der Gesellschaft der Wissenschaften, Göttingen. *Math. Phys. Kl.* **1918**, 2, 98–100
4. S.M. Londoño-Restrepo, R. Jeronimo-Cruz, B.M. Malo, E.R. Muñoz, M.R. García, Efect of the Nano Crystal Size on the X-ray Difraction Patterns of Biogenic Hydroxyapatite from Human, Bovine, and Porcine Bones. Sci. Rep. **9**, 5915 (2019). https://doi.org/10.1038/s41598-019-42269-9
5. N.A.S. Mohd Pu'ad, P. Koshy, H.Z. Abdullah, M.I. Idris, T.C. Lee, Syntheses of hydroxyapatite from natural sources. Heliyon **2019**, 5, e01588, https://doi.org/10.1016/j.heliyon.2019.e01588
6. S.J. Lee, Y.C. Lee, Y.S. Yoon, Characteristics of calcium phosphate powders synthesized from cuttlefish bone and phosphoric acid. J. Ceram. Process. Res. **8**, 427–430 (2007)
7. Manalu1, J.L.; Soegijono, B.; Indrani, D.J. Characterization of Hydroxyapatite Derived from Bovine Bone. Asian J. Appl. Sci. **2015**, 3, 758–765
8. M.K. Herliansyah, D.A. Nasution, M. Hamdi, A. Ide-Ektessabi, M.W. Wildan, A.E. Tontowi, Preparation and characterization of natural hydroxyapatite: a comparative study of bovine bone hydroxyapatite and hydroxyapatite from calcite. Mater. Sci. Forum **561–565**, 1441–1444 (2007). https://doi.org/10.4028/www.scientific.net/MSF.561-565.1441
9. M.E. Bahrololoom, M. Javidia, S. Javadpoura, J. Ma, Characterisation of natural hydroxyapatite extracted from bovine cortical bone ash. J. Ceram. Process. Res. **10**, 129–138 (2009)
10. S. Shahabi, F. Najafi, A. Majdabadi, T. Hooshmand, M. Haghbin, B. Karimi, S.M. Fatemi, Effect of gamma irradiation on structural and biological properties of a PLGA-PEG-Hydroxyapatite composite. Sci. World J. **420616**, 9 (2014). https://doi.org/10.1155/2014/420616
11. M. Markovic, B.O. Fowler, M.S. Tung, Preparation and comprehensive characterization of a calcium hydroxyapatite reference material. J. Res. Natl. Inst. Stand. Technol. **109**, 553–568 (2004). https://doi.org/10.6028/jres.109.042

12. S. Madhavi, C. Ferraris, T.J. White, Synthesis and crystallization of macroporous hydroxyapatite. J. Solid State Chem. **178**, 2838–2845 (2005). https://doi.org/10.1016/j.jssc.2005.06.039
13. A. Esmaeilkhanian, F. Sharifianjazi, A. Abouchenari, A. Rouhani, N. Parvin, M. Irani, Synthesis and Characterization of Natural Nano-hydroxyapatite Derived from Turkey Femur-Bone Waste. Appl. Biochem. Biotechnol. **189**, 919–932 (2019). https://doi.org/10.1007/s12010-019-03046-6
14. R. Das, S.S. Nath, R. Bhattacharjee, Preparation of linoleic acid capped gold nanoparticles and their spectra. Phys. E Low Dimens. Syst. Nanostruct. **43**, 224–227 (2010). https://doi.org/10.1016/j.physe.2010.07.008
15. P.C. Dey, R. Das, Effect of silver doping on the elastic properties of CdS nanoparticles. J. Indian J. Phys. **92**, 1099–1108 (2018). https://doi.org/10.1007/s12648-018-1214-4
16. M. De, S.P.S. Gupta, Lattice imperfection studies in polycrystalline materials by x-ray diffraction line-profile analysis. Pramana **23**, 721–744 (1984)
17. Y.T. Prabhu, K.V. Rao, V.S.S. Kumar, B.S. Kumari, X-Ray Analysis by Williamson-Hall and Size-Strain Plot Methods of ZnO Nanoparticles with Fuel Variation. World J. Nano Sci. Eng. **4**, 21–28 (2014). https://doi.org/10.4236/wjnse.2014.41004
18. A. Monshi, P.F. Messer, Ratio of slopes method for quantitative X-ray diffraction analysis. J. Mater. Sci. **26**, 3623–3627 (1991). https://doi.org/10.1007/BF00557154
19. A. Monshi, M.R. Foroughi, M.R. Monshi, Modified Scherrer Equation to Estimate More Accurately Nano-Crystallite Size Using XRD. World J. Nano Sci. Eng. **2**, 154–160 (2012). https://doi.org/10.4236/wjnse.2012.23020
20. D. Nath, F. Singh, R. Das, X-ray diffraction analysis by Williamson-Hall, Halder-Wagner and size-strain plot methods of CdSe nanoparticles—A comparative study. Mater. Chem. Phys. **239**, 122021 (2019). https://doi.org/10.1016/j.matchemphys.2019.122021
21. Jacob, R.; Isac, J. X-ray diffraction line profile analysis of Ba0.6Sr0.4FexTi(1-x) O3-δ, (x=0.4). Int. J. Chem. Stud. **2015**, 2, 12–21
22. B.E. Warren, B.L. Averbach, The separation of cold-work distortion and particle size broadening in X-ray patterns. J. Appl. Phys. **23**, 497 (1952). https://doi.org/10.1063/1.1702234
23. Adachi, S. Handbook on physical properties of semiconductors; Springer: New York, NY, USA, 2004, ISBN 1402077475
24. M. Bodke, U. Gawai, A. Patil, B. Dole, Estimation of accurate size, lattice strain using Williamson-Hall models, SSP and TEM of Al doped ZnO nanocrystals. Matériaux Tech. **106**, 602 (2018). https://doi.org/10.1051/mattech/2018055
25. J.M. Zhang, Y. Zhang, K.W. Xu, V. Ji, General compliance transformation relation and applications for anisotropic hexagonal metals. Solid State Commun. **139**, 87–91 (2006). https://doi.org/10.1016/j.ssc.2006.05.026
26. J.L. Katz, K. Ukraincik, On the anisotropic elastic properties of hydroxyapatite. J. Biomech. **4**, 221–227 (1971)
27. V.D. Mote, Y. Purushotham, B.N. Dole, Williamson-Hall analysis in estimation of lattice strain in nanometer-sized ZnO particles. J. Theor. Appl. Phys. **6**, 2251–7235 (2012)
28. L. Motevalizadeh, Z. Heidary, M.E. Abrishami, Facile template-free hydrothermal synthesis and microstrain measurement of ZnO nano rods. J. Bull. Mater. Sci. **37**, 397–405 (2014)
29. N.C. Halder, C.N.J. Wagner, Separation of particle size and lattice strain in integral breadth measurements. Acta Cryst. **20**, 312–331 (1966). https://doi.org/10.1107/S0365110X66000628
30. D. Balzar, H. Ledbetter, Voigt-function modeling in fourier analysis of size- and strain-broadened X-ray diffraction peaks. J. Appl. Crystal. **26**, 97–103 (1993). https://doi.org/10.1107/S0021889892008987
31. M. Mazhidi, P.H. Khani, Structural characterization of ZnO and ZnO: Mn nanoparticles prepared by reverse micelle method. Int. J. Nano Dim. **4**, 233 (2008). https://doi.org/10.7508/IJND.2011.04.004
32. S. Joscheki, B. Nies, R. Krotz, A. Goferich, Chemical and physicochemical characterization of porous hydroxyapatite ceramics made of natural bone. Biomater. **21**, 1645–1658 (2000). https://doi.org/10.1016/s0142-9612(00)00036-3

33. D.N. Misra, R.L. Bowen, G.L. Mattamal, Surface area of dental enamel, bone, and hydroxyapatite: Chemisorption from solution. J. Calcif. Tissue Res. **26**, 139–142 (1978)
34. A. Doostmohammadi, A. Monshi, R. Salehi, M.H. Fathi, S. Karbasi, U. Pieles, A.U. Daniels, Preparation, chemistry and physical properties of bone-derived hydroxyapatite particles having a negative zeta potential. Mater. Chem. Phys. **132**, 446–452 (2012). https://doi.org/10.1016/j.matchemphys.2011.11.051
35. M.J. Mayo, Processing of nanocrystalline ceramics from ultrafine particles. Int. Mater. Rev. **41**, 85–115 (1996). https://doi.org/10.1179/imr.1996.41.3.85
36. A. Ruksudjarit, K. Pengpat, G. Rujijanagul, T. Tunkasiri, Synthesis and characterization of nanocrystalline hydroxyapatite from natural bovine bone. Curr. Appl. Phys. **8**, 270–272 (2008). https://doi.org/10.1016/j.cap.2007.10.076
37. A.M. Janus, M. Faryna, K. Haberko, A. Rakowska, T. Panz, Chemical and microstructural characterization of natural hydroxyapatite derived from pig bones. Microchim. Acta **161**, 349–353 (2008). https://doi.org/10.1007/s00604-007-0864-2
38. S. Paul, A. Pal, A.R. Choudhury, S. Bodhak, V.K. Balla, A. Sinha, M. Das, Effect of trace elements on the sintering effect of fish scale derived hydroxyapatite and its bioactivity. Ceram. Int. **43**, 15678–15684 (2017). https://doi.org/10.1016/j.ceramint.2017.08.127
39. M.R. Ayatollahi, M.Y. Yahya, H.A. Shirazi, S.A. Hassan, Mechanical and tribological properties of hydroxyapatite nanoparticles extracted from natural bovine bone and the bone cement developed by nano-sized bovine hydroxyapatite filler. Ceram. Int. **41**, 10818–10827 (2015). https://doi.org/10.1016/j.ceramint.2015.05.021
40. M.R. Foroughi, S. Karbasi, R.E. Kahrizsangi, Physical and mechanical properties of a poly-3-hydroxybutyratecoated nanocrystalline hydroxyapatite scaffold for bone tissue engineering. J. Porous Mater. **19**, 667–675 (2012). https://doi.org/10.1007/s10934-011-9518-1
41. E.A. Ofudje, A. Rajendran, A.I. Adeogun, M.A. Idowu, S.O. Kareem, D. Pattanayak, Synthesis of organic derived hydroxyapatite scaffold from pig bone waste for tissue engineering applications. Adv. Powder Technol. **29**, 1–8 (2018). https://doi.org/10.1016/j.apt.2017.09.008
42. B. Mondal, S. Mondal, A. Mondal, N. Mandal, Fish scale derived hydroxyapatite scaffold for bone tissue engineering. Mater Charact **121**, 112–124 (2016). https://doi.org/10.1016/j.matchar.2016.09.034
43. A. Pal, S. Maity, S. Chabri, S. Bera, A.R. Chowdhury, M. Das, A. Sinha, Mechanochemical synthesis of nanocrystalline hydroxyapatite from Mercenaria clam shells and phosphoric acid. Biomed. Phys. Eng. Express **3**, 015010 (2017). https://doi.org/10.1088/2057-1976/aa54f5

Chapter 3
Determination of Elasticity Modulus in Relation to Atomic Density of Planar Structures in Crystal Lattice Unit Cells

This chapter exemplifies a novel technique for determining the elastic modulus of crystallite structures, such as NaCl (FCC) [1]. Young's modulus (E), a pivotal parameter in the mechanical properties of solid materials, is defined by the proportionality between stress and strain. Various experimental and theoretical methods exist to determine Young's modulus, including stress–strain curves in compression and tensile tests, electromagnetic-acoustic resonance, ultrasonic pulse-echo, and density functional theory (DFT) across different basis sets. Notably, preparing specimens for experimental measurement of Young's modulus can be cumbersome and time-consuming. Additionally, software-based calculations of Young's modulus require assumptions regarding data and structures. This chapter introduces a new method to ascertain Young's modulus of crystalline materials using X-Ray diffraction. The study calculated Young's modulus along arbitrary planes, such as random (hkl), employing the relationship between elastic compliances, crystal lattice geometry, and planar density, as determined by X-Ray diffraction. Sodium chloride (NaCl) with FCC lattice structures is used as an example. The X-Ray diffraction data, elastic stiffness constants, and elastic compliances were sourced from X'Pert software, literature, and experimental measurements, respectively. The elastic stiffness constant and Young's modulus of NaCl were also measured using ultrasonic techniques, with the results closely aligning with this new method. The modified Williamson-Hall (W–H) method in the Uniform Stress Deformation Model (USDM) is used to introduce a new approach that utilizes the W–H equation, applying a least squares technique to minimize error sources.

Supplementary Information The online version contains supplementary material available at https://doi.org/10.1007/978-3-032-12583-5_3.

M. Rabiei et al., *Hydroxyapatite-Based Nanocomposites*, Springer Series in Chemical Physics 127, https://doi.org/10.1007/978-3-032-12583-5_3

3.1 Introduction

Young's modulus is a measure of the stiffness of elastic materials, defined as the ratio of stress to strain. The response of a crystal lattice to applied forces is described by the elastic constants, including bulk modulus, shear modulus, Young's modulus, and Poisson's ratio [2]. These elastic constants are key in assessing material strength [2]. There is a relationship between the values of elastic constants and planar density, linked to the bonding characteristics between adjacent atomic planes and the anisotropic nature of bonding and structural stability [3, 4]. Elastic constants describe the stress–strain relationship and depend on the crystal lattice configuration; hence, they are derived from planes within the lattice [5–8]. Numerous studies have examined the relationship between elastic constants and particular planes/directions, as seen in Refs [9–11]. Equivalent crystallographic planes share similar atomic planar densities. Planar density, or the fraction of a crystallographic plane's area filled by atoms, is an important parameter of a crystal structure, defined as the number of atoms per unit area on a plane [12]. X-Ray diffraction (XRD) is a conventional method used to analyze materials, allowing the determination of crystalline size, stress, strain, and energy density.[13] X-ray diffraction is uniquely capable of determining both the mechanical and microstructural states for each diffracted plane. These planes function as strain gauges, enabling the measurement of Young's modulus across one or more planes/directions of the diffraction vector [14]. At present, X-ray diffraction is a standard technique for analyzing crystal structures and atomic spacing. It operates on the principle of constructive interference between monochromatic X-rays and a crystalline sample [15]. In 1969, Hanabusa et al. introduced a novel method to measure the elastic constant of the cementite phase in steel. However, this method was limited to high-angle regions, failing to detect low-angle regions [16]. Alternatively, the Williamson-Hall (W–H) method proved effective for calculating and estimating strain, providing mechanical results like lattice strain (ε), lattice stress (σ), and lattice strain energy density (u) [17]. This study introduces a new method to accurately determine Young's modulus (E) values via X-ray diffraction. To validate the practical parameters, NaCl powder was chosen for the calculations. The results closely matched experimental data and existing literature, suggesting that this new approach to calculating Young's modulus (E) using X-ray diffraction is suitable for both single-crystal and polycrystalline materials.

3.2 Materials and Experiments

In this study, a Bruker D8 Advance X-ray diffractometer with $Cu_{K\alpha}$ radiation was used. The powder X-ray diffraction was conducted at conditions of 40 kV and 40 mA, with data recorded over a range of 10–100 degrees for 2θ at a scanning speed of 2.5 degrees per minute and a step size of 0.02 degrees. The XRD patterns were analyzed using the High Score X'Pert software and the data were exported as ASC suffix files.

Merck sodium chloride powder was prepared for analysis. The pulse echo method was employed for determining the sound velocity of both transverse and longitudinal ultrasonic waves, with a specimen of approximately 2 cm thickness being fabricated for this purpose. The pulser receiver utilized was from Panametrics Co. (Waltham), and the oscilloscope employed was an Iwatsu model from Japan, with a capacity of 100 MHz. Resonant frequencies were identified as 10 MHz for longitudinal waves and 5 MHz for transverse waves. In addition, the three-dimensional geometric representation of crystal structures was created using Crystal Maker, Version 10.2.2 software.

3.3 Methods

3.3.1 X-Ray Diffraction of Compound X

In this scenario, the combination of X-Ray diffraction analysis on crystalline materials with the measurement of the planar density of each diffracted plane is conducted. This approach allows for the precise determination of the Young's modulus for each crystalline solid material. A representative XRD pattern is selected for compound x, using either a powder or crystal sample (Fig. 3.1). It is assumed that these crystallites are randomly oriented with respect to each other. Additionally, when the powder is exposed to a monochromatic X-Ray beam, diffraction occurs from planes within the crystallites that have the correct orientation to satisfy the Bragg condition. According to the Fig. 3.1, five planes consisted of ($h_1k_1l_1$), ($h_2k_2l_2$), ($h_3k_3l_3$), ($h_4k_4l_4$) and ($h_5k_5l_5$) were diffracted and diffracted beams were taken an angle 2Θ for each plane with the incident beam. As the result extracted by X'Pert software, the lattice parameter (a), index of planes (hkl) and lattice type (seven crystal systems) of compound will be recognized. For following the gaining Young's modulus value with high accuracy, the elastic stiffness constant values (C_{ij}) and elastic compliance values (S_{ij}) are needed.

3.3.2 Elastic Stiffness Constant and Elastic Compliance

The stress components are specified as the σ_{ij} that i is direction of stress and j is direction perpendicular to the plane upon which the stress is acting, the simple stress tensor brought in the Fig. 3.2. Meanwhile, σ_{xy} is a stress in the x direction acting on the plane perpendicular to the y axis. Taking into account to Fig. 3.2, it is revealed that there are nine stresses components, however, since with presumption is nor rotating, ($\sigma_{ij}= \sigma_{ji}$) the stress components are reduced to six; three axial such as σ_{xx}, σ_{yy}, σ_{zz} and three shear σ_{xy}, σ_{yz}, σ_{xz} [18].

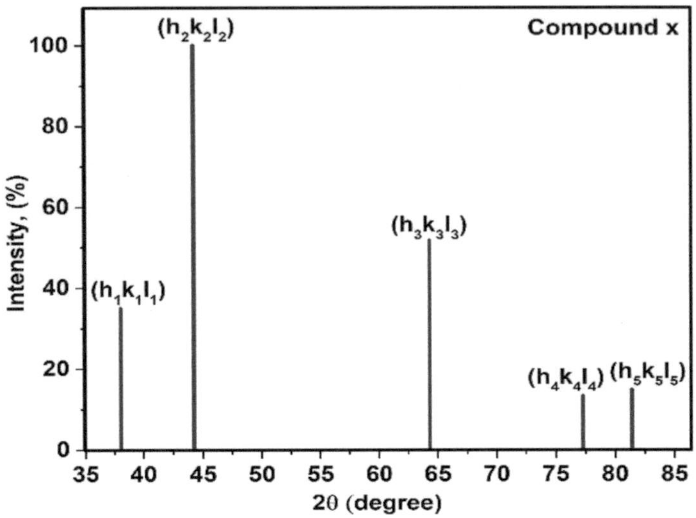

Fig. 3.1 X-ray diffraction of compound x

Fig. 3.2 Elaborates of stress

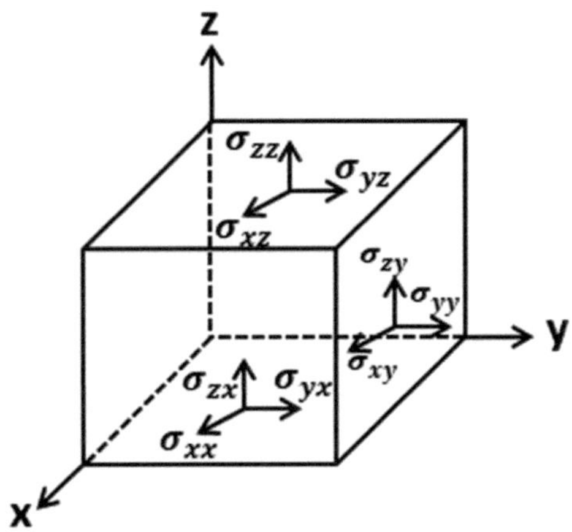

According to the Eq. 3.1 (Hooke's law), the stress is proportional to the strain for small displacements. In the generalized form, this proportionality principle is extended to the six stresses and strains [19].

$$E = \frac{\sigma}{\varepsilon} \tag{3.1}$$

Nevertheless, Hooke's law can be written such as:

$\sigma_{xx} = C_{11}\varepsilon_{xx} + C_{12}\varepsilon_{yy} + C_{13}\varepsilon_{zz} + C_{14}\varepsilon_{yz} + C_{15}\varepsilon_{zx} + C_{16}\varepsilon_{xy}$
$\sigma_{yy} = C_{21}\varepsilon_{xx} + C_{22}\varepsilon_{yy} + C_{23}\varepsilon_{zz} + C_{24}\varepsilon_{yz} + C_{25}\varepsilon_{zx} + C_{26}\varepsilon_{xy}$
$\sigma_{zz} = C_{31}\varepsilon_{xx} + C_{32}\varepsilon_{yy} + C_{33}\varepsilon_{zz} + C_{34}\varepsilon_{yz} + C_{35}\varepsilon_{zx} + C_{36}\varepsilon_{xy}$

$\sigma_{yz} = C_{41}\varepsilon_{xx} + C_{42}\varepsilon_{yy} + C_{43}\varepsilon_{zz} + C_{44}\varepsilon_{yz} + C_{45}\varepsilon_{zx} + C_{46}\varepsilon_{xy}$
$\sigma_{zx} = C_{51}\varepsilon_{xx} + C_{52}\varepsilon_{yy} + C_{53}\varepsilon_{zz} + C_{54}\varepsilon_{yz} + C_{55}\varepsilon_{zx} + C_{56}\varepsilon_{xy}$
$\sigma_{xy} = C_{61}\varepsilon_{xx} + C_{62}\varepsilon_{yy} + C_{63}\varepsilon_{zz} + C_{64}\varepsilon_{yz} + C_{65}\varepsilon_{zx} + C_{66}\varepsilon_{xy}$

The constants of proportionality C_{ij} are introduced as the elastic constants. Because for small displacement the elastic energy density is a quadratic function of the strains, and the elastic constants are given by the second partial derivatives of energy density, so it can be shown $C_{ij}= C_{ji}$. Meanwhile, the 36 elastic constants are resulted in the matrix form and they can be decreased to 6 diagonal components and 15 off-diagonal elaborates. In addition, it is possible for each system of crystal structures, that the number of matrix can be decreased. According to the Eq. 3.2 in matrix form, these elastic constants can be written as [20, 21]:

$$\begin{vmatrix} C_{11} & C_{12} & C_{13} & C_{14} & C_{15} & C_{16} \\ C_{21} & C_{22} & C_{23} & C_{24} & C_{25} & C_{26} \\ C_{31} & C_{32} & C_{33} & C_{34} & C_{35} & C_{36} \\ C_{41} & C_{42} & C_{43} & C_{44} & C_{45} & C_{46} \\ C_{51} & C_{52} & C_{53} & C_{54} & C_{55} & C_{56} \\ C_{61} & C_{62} & C_{63} & C_{64} & C_{65} & C_{66} \end{vmatrix} \tag{3.2}$$

The specified form of Eq. 3.1 can be written in decreased tensor notation and the form was so called Voigt notation (Eq. 3.3).

$$\sigma_i = C_{ij}\varepsilon_j \tag{3.3}$$

i and j equals 1–6, and i,j which goes from 1 through 6, related to the xx, yy, zz, yz, zx, and xy, respectively.

Furthermore, in the practical subjects, it is preferred taking into account to strains in terms of the stresses. According to the Eqs. 3, 4 can be written

$$\varepsilon_i = S_{ij}\sigma_j \tag{3.4}$$

that, the S_{ij} is introduced as the elastic compliance constants, so the exact formula can be written as the below

$\varepsilon_1 = S_{11}\sigma_1 + S_{12}\sigma_2 + S_{13}\sigma_3 + S_{14}\sigma_4 + S_{15}\sigma_5 + S_{16}\sigma_6$
$\varepsilon_2 = S_{21}\sigma_1 + S_{22}\sigma_2 + S_{23}\sigma_3 + S_{24}\sigma_4 + S_{25}\sigma_5 + S_{26}\sigma_6$
$\varepsilon_3 = S_{31}\sigma_1 + S_{32}\sigma_2 + S_{33}\sigma_3 + S_{34}\sigma_4 + S_{35}\sigma_5 + S_{36}\sigma_6$
$\varepsilon_4 = S_{41}\sigma_1 + S_{42}\sigma_2 + S_{43}\sigma_3 + S_{44}\sigma_4 + S_{45}\sigma_5 + S_{46}\sigma_6$
$\varepsilon_5 = S_{51}\sigma_1 + S_{52}\sigma_2 + S_{53}\sigma_3 + S_{54}\sigma_4 + S_{55}\sigma_5 + S_{56}\sigma_6$
$\varepsilon_6 = S_{61}\sigma_1 + S_{62}\sigma_2 + S_{63}\sigma_3 + S_{64}\sigma_4 + S_{65}\sigma_5 + S_{66}\sigma_6$

The seven classic unit cells of Bravais lattices comprise Triclinic, Monoclinic, Orthorhombic, Tetragonal, Cubic, Trigonal, and Hexagonal. Each of these lattices possesses distinct values for both elastic stiffness and compliance. Conventionally, the Triclinic matrix is understood to have 21 elastic constants. However, as Landau and Lifshitz argued, this figure should be reduced to 18. This reduction is attributed to the arbitrary selection and interchangeability of the Cartesian orthogonal axes x, y, and z, which effectively reduces the count by three [22].

$$\text{Triclinic matrix}\begin{vmatrix} C_{11} & C_{12} & C_{13} & C_{14} & C_{15} & C_{16} \\ C_{21} & C_{22} & C_{23} & C_{24} & C_{25} & C_{26} \\ C_{31} & C_{32} & C_{33} & C_{34} & C_{35} & C_{36} \\ C_{41} & C_{42} & C_{43} & C_{44} & C_{45} & C_{46} \\ C_{51} & C_{52} & C_{53} & C_{54} & C_{55} & C_{56} \\ C_{61} & C_{62} & C_{63} & C_{64} & C_{65} & C_{66} \end{vmatrix}$$

Monoclinic has 13 elastic constants and due to the suitable selection of the coordinate axes, these should be decreased to 12.

$$\text{Monoclinic matrix}\begin{vmatrix} C_{11} & C_{12} & C_{13} & 0 & C_{15} & 0 \\ C_{21} & C_{22} & C_{23} & 0 & C_{25} & 0 \\ C_{31} & C_{32} & C_{33} & 0 & C_{35} & 0 \\ 0 & 0 & 0 & C_{44} & 0 & C_{46} \\ C_{51} & C_{52} & C_{53} & 0 & C_{55} & 0 \\ 0 & 0 & 0 & C_{64} & 0 & C_{66} \end{vmatrix}$$

Elastic constants orthorhombic are obtained from nine components.

$$\text{Orthorhombic matrix}\begin{vmatrix} C_{11} & C_{12} & C_{13} & 0 & 0 & 0 \\ C_{21} & C_{22} & C_{23} & 0 & 0 & 0 \\ C_{31} & C_{32} & C_{33} & 0 & 0 & 0 \\ 0 & 0 & 0 & C_{44} & 0 & 0 \\ 0 & 0 & 0 & 0 & C_{55} & 0 \\ 0 & 0 & 0 & 0 & 0 & C_{66} \end{vmatrix}$$

Trigonal has two types of elastic constants, first one for the symmetry groups that they have seven constants and second one for suitable selection of coordinates, these should be decreased to six constants.

$$\text{Trigonal matrix (1)}\begin{vmatrix} C_{11} & C_{12} & C_{13} & C_{14} & -C_{25} & 0 \\ C_{12} & C_{11} & C_{13} & -C_{14} & C_{25} & 0 \\ C_{13} & C_{13} & C_{33} & 0 & 0 & 0 \\ C_{14} & -C_{14} & 0 & C_{44} & 0 & C_{25} \\ -C_{25} & C_{25} & 0 & 0 & C_{44} & C_{14} \\ 0 & 0 & 0 & C_{25} & C_{14} & \frac{1}{2}(C_{11}-C_{12}) \end{vmatrix}$$

$$\text{Trigonal matrix (2)}\begin{vmatrix} C_{11} & C_{12} & C_{13} & C_{14} & 0 & 0 \\ C_{12} & C_{11} & C_{13} & -C_{14} & 0 & 0 \\ C_{13} & C_{13} & C_{33} & 0 & 0 & 0 \\ C_{14} & -C_{14} & 0 & C_{44} & 0 & C_{25} \\ -C_{25} & C_{25} & 0 & 0 & C_{44} & C_{14} \\ 0 & 0 & 0 & 0 & C_{14} & \frac{1}{2}(C_{11}-C_{12}) \end{vmatrix}$$

Tetragonal system has two types of elastic constants as well, first one for the symmetry groups that they have seven constants and second one for suitable selection of coordinates, these should be decreased to six constants [21, 23].

$$\text{Tetragonal matrix (1)}\begin{vmatrix} C_{11} & C_{12} & C_{13} & 0 & 0 & C_{16} \\ C_{12} & C_{11} & C_{13} & 0 & 0 & -C_{16} \\ C_{13} & C_{13} & C_{33} & 0 & 0 & 0 \\ 0 & 0 & 0 & C_{44} & 0 & 0 \\ 0 & 0 & 0 & 0 & C_{44} & 0 \\ C_{16} & -C_{16} & 0 & 0 & 0 & C_{66} \end{vmatrix}$$

$$\text{Tetragonal matrix (2)}\begin{vmatrix} C_{11} & C_{12} & C_{13} & 0 & 0 & 0 \\ C_{12} & C_{11} & C_{13} & 0 & 0 & 0 \\ C_{13} & C_{13} & C_{33} & 0 & 0 & 0 \\ 0 & 0 & 0 & C_{44} & 0 & 0 \\ 0 & 0 & 0 & 0 & C_{44} & 0 \\ 0 & 0 & 0 & 0 & 0 & C_{66} \end{vmatrix}$$

The hexagonal system has five elastic constants

$$\text{Hexagonal matrix}\begin{vmatrix} C_{11} & C_{12} & C_{13} & 0 & 0 & 0 \\ C_{12} & C_{11} & C_{13} & 0 & 0 & 0 \\ C_{13} & C_{13} & C_{33} & 0 & 0 & 0 \\ 0 & 0 & 0 & C_{44} & 0 & 0 \\ 0 & 0 & 0 & 0 & C_{44} & 0 \\ 0 & 0 & 0 & 0 & 0 & \frac{1}{2}(C_{11} - C_{12}) \end{vmatrix}$$

The Cubic system has three elastic constants

$$\text{Cubic matrix}\begin{vmatrix} C_{11} & C_{12} & C_{13} & 0 & 0 & 0 \\ C_{12} & C_{11} & C_{13} & 0 & 0 & 0 \\ C_{13} & C_{13} & C_{33} & 0 & 0 & 0 \\ 0 & 0 & 0 & C_{44} & 0 & 0 \\ 0 & 0 & 0 & 0 & C_{44} & 0 \\ 0 & 0 & 0 & 0 & 0 & C_{44} \end{vmatrix}$$

For conventional systems consisted of cubic and hexagonal structures the relationship between C_{ij} and S_{ij} are presented in Eq. 3.5–3.12 respectively.

For Cubic;

$$S_{11} = \frac{C_{11} + C_{12}}{(C_{11} - C_{12})(C_{11} + 2C_{12})} \text{ [25]} \tag{3.5}$$

$$S_{12} = \frac{-C_{12}}{(C_{11} - C_{12})(C_{11} + 2C_{12})} \text{ [25]} \tag{3.6}$$

$$S_{44} = \frac{1}{C_{44}} \text{ [25]} \tag{3.7}$$

For Hexagonal;

$$S_{11} = \frac{1}{2}\left(\frac{C_{33}}{C_{33}(C_{11} + C_{12}) - 2(C_{13})^2} + \frac{1}{C_{11} - C_{12}}\right) \text{ [24]} \tag{3.8}$$

$$S_{12} = \frac{1}{2}\left(\frac{C_{33}}{C_{33}(C_{11}+C_{12})-2(C_{13})^2} - \frac{1}{C_{11}-C_{12}}\right) [26] \tag{3.9}$$

$$S_{33} = \frac{C_{11}+C_{12}}{C_{33}(C_{11}+C_{12})-2(C_{13})^2} [26] \tag{3.10}$$

$$S_{13} = -\frac{C_{13}}{C_{33}(C_{11}+C_{12})-2(C_{13})^2} [26] \tag{3.11}$$

$$S_{44} = \frac{1}{C_{44}} [26] \tag{3.12}$$

Mostly, elastic constant values of materials are tabulated in the literatures [25, 26]. Furthermore, Young's modulus of each plane (E_{hkl}) can be expressed for cubic and hexagonal crystals respectively as the (Eqs. 3.13 and 3.14).

For Cubic;

$$\frac{1}{E_{hkl}} S_{11} - 2\left[(S_{11}-S_{12}) - \frac{1}{2}S_{44}\right] \cdot \left[\frac{h^2k^2 + k^2l^2 + l^2h^2}{(h^2+k^2+l^2)}\right] [27] \tag{3.13}$$

For Hexagonal;

$$E_{hkl} = \frac{\left[h^2 + \frac{(h+2k)^2}{3} + \left(\frac{al}{c}\right)^2\right]^2}{S_{11}\left(h^2 + \frac{(h+2k)^2}{3}\right)^2 + S_{33}\left(\frac{al}{c}\right)^4 + (2S_{13}+S_{44})\left(h^2 + \frac{(h+2k)^2}{3}\right)\left(\frac{al}{c}\right)^2} [28, 29] \tag{3.14}$$

3.3.3 Relationship Between Young's Modulus E_{hkl} and Planar Density of Each Diffracted Plane Through the X-Ray Diffraction

Type of crystal lattice of compound x must be determined according to the X-Ray diffraction file. In this case presumed that the compound x has cubic crystal structure. According to the Eqs. 3.5–3.7, elastic compliance values of compound x were calculated as the S_{11}, S_{12} and S_{44}, therefore, Young's modulus values of diffracted planes (Fig. 3.1) were registered by Eq. 3.13. The values of Young's modulus of diffracted planes or modulus of Elasticity (E) of compound x named $E_{(h_1k_1l_1)}$, $E_{(h_2k_2l_2)}$, $E_{(h_3k_3l_3)}$, $E_{(h_4k_4l_4)}$ and $E_{(h_5k_5l_5)}$ respectively. In addition, planar Density (PD) values of diffracted planes were calculated as the $PD_{(h_1k_1l_1)}$, $PD_{(h_2k_2l_2)}$, $PD_{(h_3k_3l_3)}$, $PD_{(h_4k_4l_4)}$ and $PD_{(h_5k_5l_5)}$. Applying least squares method for the line between values extracted of Young's modulus and planar density

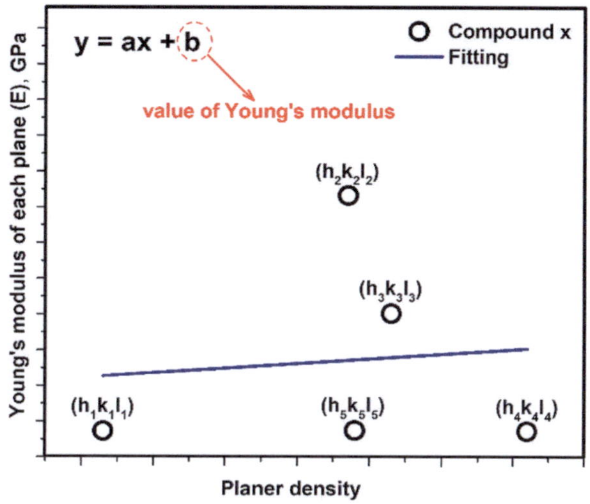

Fig. 3.3 Young's modulus extracted of planes of compound x versus planar density

of diffracted planes can give the average Young's modulus value of compound x with high accuracy. In this method Young's modulus values of diffracted planes played role as a y axis, and x axis is planar density. Young's modules value extracted of each plane of compound x versus planar density are presented in Fig. 3.3.

In this method, the unit of Young's modulus value is depended to the unit of elastic stiffness constant and elastic compliance values, and here is GPa. Moreover, planar density doesn't have unit always and the value of planar density is less than 1 always. In addition, the values of planar density are depended on the situation of planes. For example in this curve (Fig. 3.3) for compound x, $PD_{(h_4k_4l_4)} > PD_{(h_3k_3l_3)} > PD_{(h_5k_5l_5)} > PD_{(h_2k_2l_2)} > PD_{(h_1k_1l_1)}$. Fourthermore, it is possible that two or more planes have similar planar density values.

In this method, the empty planes in one unite cell are not considered in calculations. However, if there is no atom in the plane inside the unite cell, they will appear when the plane is extended to the adjacent cells and atoms will be appeared and unit cell will be converted to the super cells, and calculating planar density for two or more adjacent unit cell or super cells is possible.

In some cases such as plane (310) in simple cubic, as observed in Fig. 3.4, there are two corner atoms in the original crystal lattice. But when unit cell is extended there is no atom in the adjacent crystal. In this case the planar density should be calculated taking into account the three adjacent crystal lattices. In other words, when (310) plane is continued and extended to 3 unit cells, there is no atom in the middle one, but there are $2 \times \frac{1}{2}$ atoms in the first and third unit cells.

When atoms are compacted, higher modulus of elastisity is obtained. For example for Fe (BCC), $E_{(100)}$ is equal to 125 GPa and $E_{(111)}$ is equal to 272.7 GPa [30], since density in (111) plane of Fe (BCC) is higher and needs more force for displacement of atoms. Furthermore, controlling the process of deformation and displacement of atoms in the planes is related to the dislocation networks. The force (W) for moving

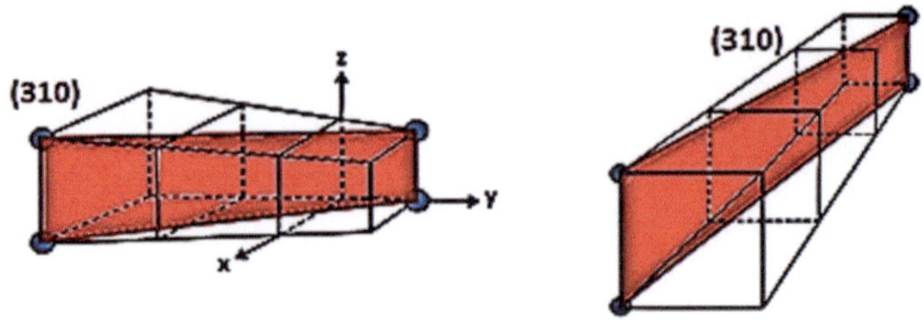

Fig. 3.4 Schematics of (310) expanded in three unit cells

atoms in each plane is corresponded to the Eq. 3.15 [31, 32].

$$W = \mathrm{Gb}^2 1 \tag{3.15}$$

where G is the shear modulus, b the Burgers vector and l is dislocation length. In addition, substitution of G with Eq. 3.16 can be provided to gaining Eq. 3.17. The E is Young's modulus and ν is Poisson's ratio.

$$\mathrm{G} = \frac{\mathrm{E}}{2(1+\nu)} \tag{3.16}$$

$$W = \frac{\mathrm{E}}{2(1+\nu)} b^2 1 \tag{3.17}$$

When atoms are at far distances, such as corner atoms in (100), atoms need higher values of applied force for approaching [33, 34]. Active slip systems, a combination of slip plane and slip direction are required for plastic deformation [34–37]. Moreover, dislocation interactions can be resulted [38]. X-Ray diffraction provides information about diffracted phase and the position of atoms in the phase. In this research the density of atoms in any plane is introduced as the phase density, which is defined as the area of the atoms with the centre located at the plane, divided to the total area of the plane, it is responsible for the mechanical properties of each plane. The fundamental assumption is that when the density of plane is increased, the movement of atoms with the mechanism of dislocation motion requires high forces. The increase of the force means that the modulus of elasticity (Young's modulus) would be get higher value. Now, we consider the modulus of elasticity for each phase with specific phase density, the useful information is obtained by plotting the Young's modulus versus planar density, by least squares method. Decreasing the sources of error by segregation of data provides the best phase and most accurate intercept. The intercept is responsible for the average Young's modulus of materials, from which mechanical properties of different materials can be specified. In the other aspect, C_{11} is in the agreement with the longitudinal distortion and longitudinal compression/tension, thus, C_{11} can be described as the hardness. Moreover, the transverse distortion is related to the C_{12}, and C_{12}, derived from the transverse expansion corresponded to

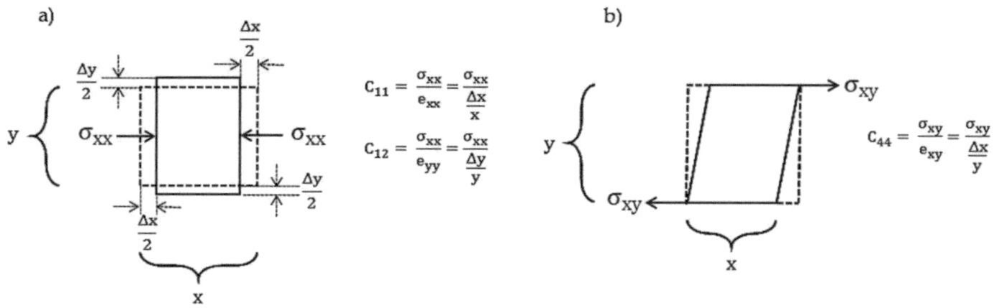

Fig. 3.5 **a** longitudinal compression C_{11} and transverse expansion C_{12} **b** shear modulus C_{44} [39]

the Poisson's ratio. In addition, C_{44} is based on the shear modulus, as well as C_{44} is in the settlement with C_{11} and C_{12} [24, 39]. The schematic role of elastic constants in cubic structures are presented in Fig. 3.5 [39]. Accordingly, shear modulus is in the relationship with Burgers vector and Young's modulus (Eq. 3.16 and 3.17), as well as, dislocation density is in the agreement with Young's modulus through the Eqs. 3.18 and 3.19 [41, 42].

$$\rho = \frac{\mathrm{E}}{\vdots\mathrm{E}} \tag{3.18}$$

where, E is Young's modulus and $\vdots\mathrm{E}$ is introduced as an Eq. 3.19.

$$\vdots\mathrm{E} = \frac{\mathrm{Gb}^2}{2} \tag{3.19}$$

Nevertheless, least squares technique can be applied for recording Young's modulus and planar density values of each diffracted plane via X-Ray diffraction. In addition, this method can be utilized for modifying Williamson-Hall (W–H) method in uniform stress deformation model (USDM) due to the creating minimize the sources of errors.

3.4 Modified W–H (USDM Model)

W–H method has used the full width at half maximum (FWHM) of the diffraction peaks for determining different elastic properties. We must decrease the errors mathematically and obtain the average value of E through the all peaks (or diffracted peaks) by using least squares method. According to the physical broadening (β_{hkl}) of X-Ray diffraction peak, it is a combination of size and strain. *W–H method is a simplified integral expanse where strain-induced broadening is distinguished through the taking into account to the peak width* [42]. *In W–H method through the USDM model,* Young's modulus values are considered. According to the Eq. 3.20, terms of $\frac{4sin\theta}{E_{hkl}}$ along X-axis and $\beta_{hkl}.cos\theta$ along Y-axis are related to the XRD pattern of samples.

$$\beta_{hkl}cos\theta = \left(\frac{K\lambda}{L}\right) - 4\sigma\frac{sin\theta}{E_{hkl}} \tag{3.20}$$

β_{hkl} introduced the broadening of peak derived from (hkl) plane. In this model the state of calculating strain and average Young's modulus values are performed. In previous research, the mean E value used in the W–H method calculations was prone to errors because relying on average E values led to deviations from the standard E values for each peak as determined by X-ray diffraction. Additionally, some studies have incorrectly used tabulated material values for E. For instance, Ref [43] calculated the average E value for cadmium selenide (CdSe) to be approximately 79.233 GPa, which significantly deviates from the standard E value reported in Refs [44, 45]. For zinc oxide (ZnO), the USDM model yielded an E value of around 130 GPa, differing from other studies cited in Refs [46, 47]. Another study utilizing the USDM model for yttrium oxide (Y2O3) [48] reported an E value inconsistent with findings in Refs [49, 50]. Similarly, in the case of sodium chloride (NaCl), elastic constant values were assessed using resonant-ultrasound spectroscopy in accordance with Refs [51, 52]. Tables 3.1 and 3.2 detail the elastic and elastic compliance constants obtained in both experimental and theoretical frameworks from earlier studies and the current investigation.

Table 3.1 Elastic constant values of NaCl extracted by experimental/theoretical literatures and this study

Elastic constant (C), (Gpa)	Expt. from Ref (Bartels et al.) [1]	Expt. from Ref (Barsch et al.) [2]	Expt. from Ref (Charles et al.) [3]	Theory. from Ref (Anderson et al.) [4]	This Study
C_{11}	48.99	49.00	50.00	49.50	49.11
C_{12}	12.57	12.60	12.70	13.20	12.26
C_{44}	12.72	12.70	14.40	12.79	13.73

Table 3.2 Elastic compliance values of NaCl extracted by experimental/theoretical literatures and this study

Elastic compliances (S), (Gpa)	Expt. by (Bartels et al.)	Expt. by (Barsch et al.)	Expt. by (Charles et al.)	Theory. by (Anderson et al.)	This Study
S_{11}	0.0228	0.0228	0.0222	0.0227	0.0226
S_{12}	– 0.0046	– 0.0046	– 0.0045	– 0.0047	– 0.0045
S_{44}	0.0786	0.0787	0.0694	0.0781	0.0728

X-Ray diffraction of NaCl is presented in Fig. 3.6. The characterization peaks of the NaCl are very close to the corresponding report in Ref [57]. NaCl has the ionic bonds, the ionic radius of Na^+ and Cl^- are 0.97 and 1.81 Å tandemly. The crystal of NaCl is FCC, the location of atom of Cl introduces (000) and Na at $(\frac{1}{2}\frac{1}{2}\frac{1}{2})$ position. According to the X'Pert analysis, lattice parameter has gained as 5.640 Å and it is in the well agreement with the values reported in the Ref [58]. In addition, crystallographic parameters of NaCl resulted by X'Pert are submitted in Table 3.3.

As illustrated in Fig. 3.6, diffraction has been observed in ten planes. Through the method outlined in this study aimed at calculating the planar density values for each diffracted plane, the schematic representation of the planes, including the arrangement of diffracted planes and the positioning of atoms in the unit cell, is provided. The configuration of the diffracted planes and atoms in NaCl, along with the planar density calculations for each diffracted plane, are depicted in Figs. 3.7 and S8.1.1.

According to the Eq. 3.13 (because NaCl has FCC structure), the values of elastic compliances (S) (Table 3.2) were substituted and E_{hkl} values of each diffrected plane were calculated as the 34.72, 44.24, 36.69, 39.18, 34.72, 36.10, 39.09, 36.69 and

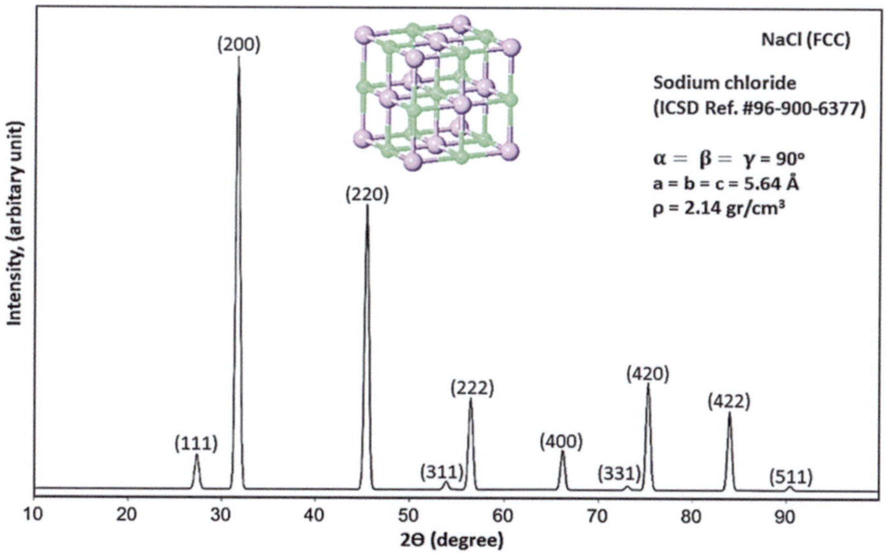

Fig. 3.6 X-ray diffraction of NaCl powder sample

Table 3.3 Crystallographic parameters of NaCl (FCC) structure resulted via X'Pert software

NaCl					
Crystal system	a (Å)	c (Å)	Cell volume (Å) ^{3}	Crystal density (g/ cm^3)	Space group
FCC	5.640	5.640	181.511	2.141	Fm-3m

a)

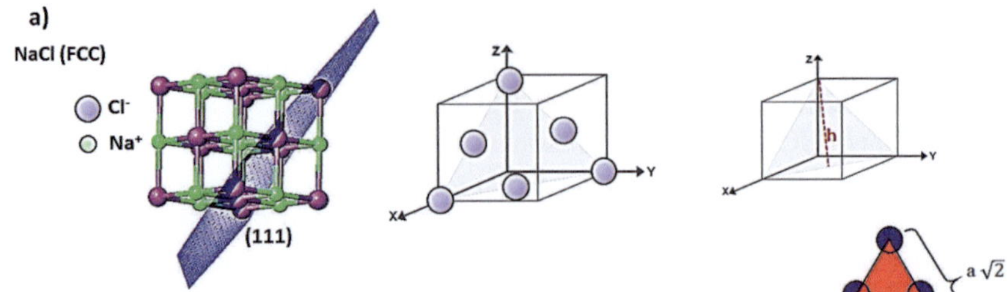

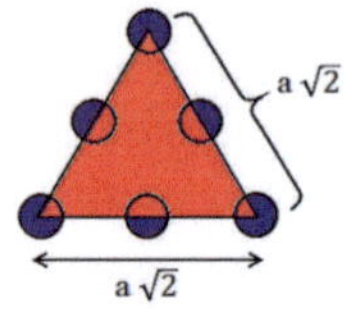

for $NaCl_{FCC}$: (ionic radius) $r_{Na^+} = 0.97$ Å, $r_{Cl^-} = 1.81$ Å, $a = 5.64$ Å

$(a\sqrt{2})^2 = (h)^2 + (\frac{a\sqrt{2}}{2})^2 \longrightarrow (7.95)^2 = (h)^2 + (3.97)^2 \longrightarrow h = 6.88$

area of the plane (111): $S = \frac{\text{height} \times \text{width}}{2} = \frac{(h) \times a\sqrt{2}}{2} = \frac{6.88 \times 7.95}{2} = 27.34$

number of atoms in the plane (111) × area of each atom in the plane (111) $= \left[\left(3 \times \frac{1}{6}\right) + \left(3 \times \frac{1}{2}\right)\right] \times \pi(r_{Cl^-_{Al}})^2 = \left[\left(3 \times \frac{1}{6}\right) \times \pi\,(1.81)^2 + \left(3 \times \frac{1}{2}\right) \times \pi\,(1.81)^2\right] = 20.57$

planer density $= \frac{\text{number of atoms in the plane (111)} \times \text{area of each atom in the plane (111)}}{\text{area of the plane (111)}} = \frac{20.57}{27.34} = 0.75$

b)

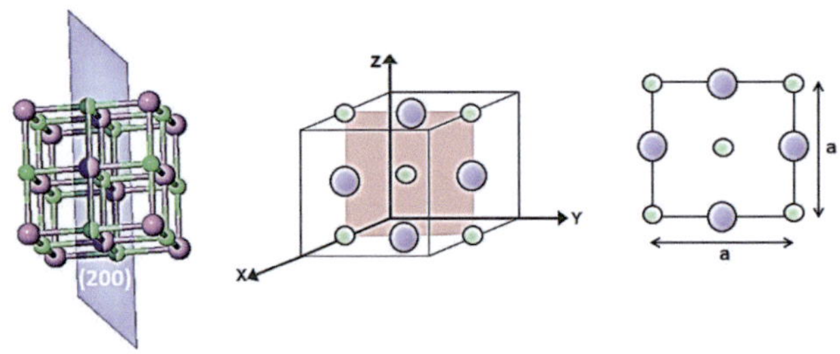

area of the plane (200): $S = a^2 = 31.80$

number of atoms in the plane (200) × area of each atom in the plane (200) =

$\left[\left(4 \times \frac{1}{4} \times \pi\,(r_{Na^+})^2\right) + \left(4 \times \frac{1}{2} \times \pi\,(r_{Cl^-})^2\right) + \left(1 \times \pi\,(r_{Na^+})^2\right)\right] =$

$\left[\left((4 \times \frac{1}{4}) \times \pi\,(0.97)^2\right) + \left(4 \times \frac{1}{2} \times \pi\,(1.81)^2\right) + (1 \times \pi\,(0.97)^2)\right] = 2.95 + 20.57 + 2.95 = 26.47$

planer density $= \frac{\text{number of atoms in the plane (200)} \times \text{area of each atom in the plane (200)}}{\text{area of the plane (200)}} = \frac{26.47}{31.80} = 0.83$

c)

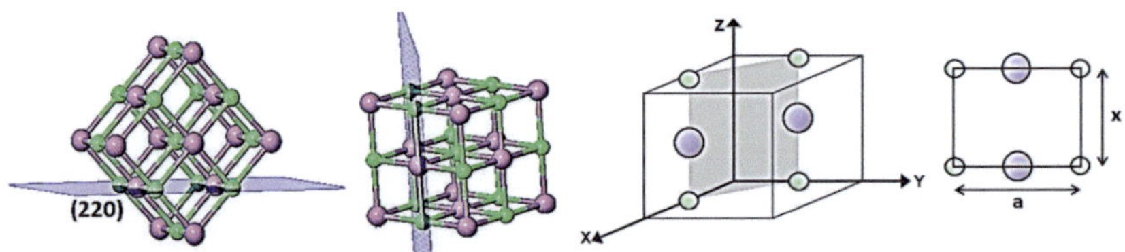

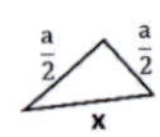

$(x)^2 = \left(\frac{a}{2}\right)^2 + \left(\frac{a}{2}\right)^2 = \left(\frac{5.64}{2}\right)^2 + \left(\frac{5.64}{2}\right)^2 = 15.90 \longrightarrow x = 3.98$

area of the plane (220): $S = a \times x = 22.44$

number of atoms in the plane (220) × area of each atom in the plane (220) $= \left(4 \times \frac{1}{4}\right) \times \pi(r_{Na^+})^2 + \left(2 \times \frac{1}{2}\right) \times \pi(r_{Cl^-})^2 =$
$(1) \times \pi(0.97)^2 + (1) \times \pi(1.81)^2 = 2.95 + 10.28 = 13.23$

planer density $= \frac{\text{number of atoms in the plane (220)} \times \text{area of each atom in the plane (220)}}{\text{area of the plane (220)}} = \frac{13.23}{22.44} = 0.58$

Fig. 3.7 Geometry and the situation of involved atoms in diffracted planes **a** (111), **b** (200) and **c** (220)

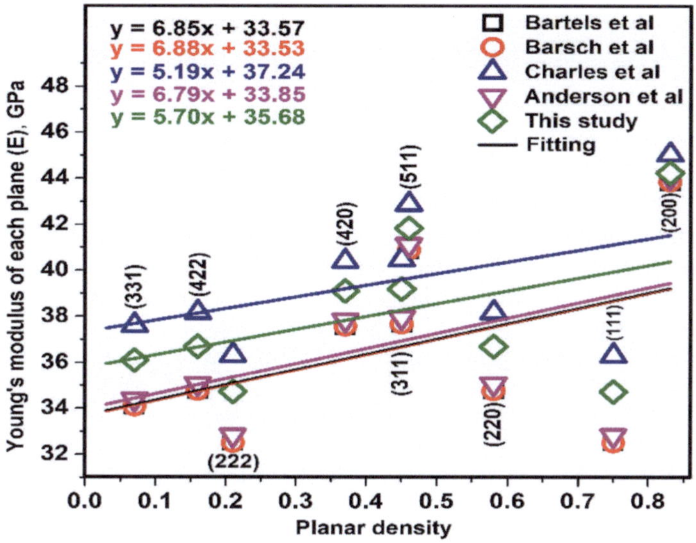

Fig. 3.8 Young's modulus of each plane of NaCl extracted by XRD patterns versus planar density

41.83 GPa for planes (111), (200), (220), (311), (222), (331), (420), (422) and (511) respectively. In addition, there is no atom in (400) (discussed in previous part), therefore this plane is not considered in the calculations. Young's modulus of each plane of NaCl extracted by XRD patterns versus planar density are presented in Fig. 3.8. Linear equations of NaCl recorded y = 6.85x + 33.57 (Bartels et al.), y = 6.88x + 33.53 (Barsch et al.), y = 5.19x + 37.24 (Charles et al.), y = 6.79x + 33.85 (Anderson et al.) and y = 5.70x + 35.68 (this study). Nevertheless, the intercepts were calculated as 33.57, 33.53, 37.24, 33.85 and 35.68 GPa respectively, the values of intercepts (Young's modulus) are submitted in Table 3.4.

$$E = \frac{\rho c_1^2\left[3\left(\frac{c_1}{c_t}\right)^2 - 4\right]}{\left(\frac{c_1}{c_t}\right)^2 - 1} \tag{3.21}$$

In this equation, ρ, c_l and c_t are density, velocity of ultrasound longitudinal and transverse waves respectively. According to the differences between two echo ($t = t_2 - t_1$) in the signals and with knowing the length of specimen, velocity of ultrasound longitudinal and transverse waves can be calculated through the Eq. 3.22.

$$c = \frac{2L}{t} \tag{3.22}$$

where, L is length of specimen and t is differences between two echoes, as well as density of specimen can be registered via measuring mass and volume of sample [63].

In addition, with placing Eq. 3.22 in the Eq. 3.21, the main formula for calculating Young's modulus value is introduced as the Eq. 3.23.

$$E = \frac{4\rho\left(\frac{L}{t_s}\right)^2\left(3t_s^2 - 4t_l^2\right)}{t_s^2 - t_l^2} \tag{3.23}$$

where, t_s and t_l are differences between two echoes in longitudinal and transverse waves respectively [61]. After measuring five times for adjustments of longitudinal and transverse waves the Young's modulus value is gained as 34.83 GPa. The value of Young's modulus extracted through the ultrasonic technique was in the well agreement with value extracted in this study (Table 3.4).

3.5 Modified Williamson-Hall Method (USDM) for NaCl

For a more realistic NaCl crystal system, the anisotropic nature of Young's modulus is considered [62]. The generalization of Hook's law (Eq. 3.1), which the strain (ε) and stress (σ) are in a linear relationship with the constant of proportionality being the modulus of elasticity or simply Young modulus. In this method, for NaCl structure, Hooke's law was performed for strain and stress, taking into account the linear proportionality to the Eq. 3.20. This equation is just an access that it is credible for a small strain notably. Furthermore, increasing the strain will be caused to the deviation of atoms from linear relationship [63]. The exact parameters resulted by X-Ray diffraction are given in Table 3.5. According to the W–H method in USDM model $\frac{4sin\theta}{E}$ (degree/GPa) played role as X axis and β(radian), cosΘ (degree) played role as Y axis (Fig. 3.9). According to the Fig. 3.9, the slope values are related to the stress (σ). Values of stress and strain have gained negative. The positive values of intrinsic strain and stress can be proving tensile strain and stress and if values were negative such as Table 3.5, they will be related to the compressive stress and strain. Moreover, stress (σ) and strain (ε) values extracted by modified W–H method in USDM model, shear modulus (μ), Poisson's ratio (υ) and bulk modulus (B) of NaCl are reported in the Table 3.6. The values of mechanical properties (Table 3.6) were in the well agreement with values tabulated in the Ref [64].

- $H = 2C_{44} + C_{12} - C_{11}$ [65]
- Shear modulus: $\mu = C_{44} - \frac{1}{5}H$ [66] [66]
- Poisson's ratio: $\upsilon = \frac{C_{12-\frac{H}{5}}}{2(C_{12}+C_{44}-2\frac{H}{5})}$ [65]
- Bulk modulus: $B = \frac{C_{11}+2C_{12}}{3}$ [67]

Table 3.4 Young's modulus values of NaCl

Study	Young modulus (E), (Gpa) in this method (intercept value)
Expt. by (Bartels et al.)	33.57
Expt. by (Barsch et al.)	33.53
Expt. by (Charles et al.)	37.24
Theory. by (Anderson et al.)	33.85
This study	35.68

To assess this approach, the ultrasonic method was employed to determine Young's modulus of NaCl. Ultrasonic waves, which are elastic waves with higher frequencies, travel through a medium to obtain Young's modulus of samples. This technique involves the use of both longitudinal and transverse waves for the measurement[59, 60]. By measuring the velocity of the ultrasound waves and the specimen'sdensity, Young's modulus is calculated (see Eq. 3.21)

3.6 Conclusion

A novel technique for precisely determining the modulus of elasticity in crystalline materials has been successfully introduced. The planar density, defined as the total count of atoms or ions per unit area of a plane, influences the plane's modulus of elasticity. By plotting the modulus of elasticity of each plane (y-axis) against its planar density (x-axis) using the least squares method, the Young's modulus of the materials can be identified at the intercept. A case study using NaCl has confirmed the accuracy of this technique, aligning well with ultrasonic methods. The Williamson-Hall method, particularly the uniform stress deformation model (USDM), can be applied within this context to reduce errors inherent in the least squares method, thus providing a more accurate elasticity modulus than average values. This technique is limited by the requirement for XRD data for planar density calculations, which assumes a uniform distribution of atoms within a crystal lattice with a unit cell, rendering it unsuitable for amorphous materials. This method has applications in both scientific research and industrial contexts.

Table 3.5 Crystallographic parameters of each individual XRD pattern related to the NaCl

NaCl										
2Θ (degree)	β = FWHM (degree)	Θ (degree)	cosΘ (degree)	1/cosΘ (degree)	Ln(1/cosΘ) (degree)	β = FWHM (radian)	Ln β (radian)	4 sinΘ (degree)	β(radian).cosΘ (degree)	hkl
27.01	0.70	13.50	0.97	1.03093	0.03046	0.01218	−4.40796	0.92	0.01181	111
30.91	0.90	15.45	0.96	1.04167	0.04082	0.01566	−4.15665	1.04	0.01503	200
45.08	0.89	22.54	0.92	1.08696	0.08338	0.01549	−4.16782	1.52	0.01425	220
53.70	0.91	26.85	0.89	1.1236	0.11653	0.01583	−4.1456	1.80	0.01409	311
56.79	0.90	28.39	0.87	1.14943	0.13926	0.01566	−4.15665	1.88	0.01362	222
66.98	0.80	33.49	0.83	1.20482	0.18633	0.01392	−4.27443	2.20	0.01155	400
72.99	0.10	36.49	0.80	1.25	0.22314	0.00174	−6.35387	2.36	0.00139	331
76.05	0.70	38.02	0.78	1.28205	0.24846	0.01218	−4.40796	2.44	0.0095	420
83.93	0.81	41.96	0.74	1.35135	0.30111	0.01409	−4.26201	2.64	0.01043	422
92.60	0.10	46.30	0.69	1.44928	0.37106	0.00174	−6.35387	2.88	0.0012	511

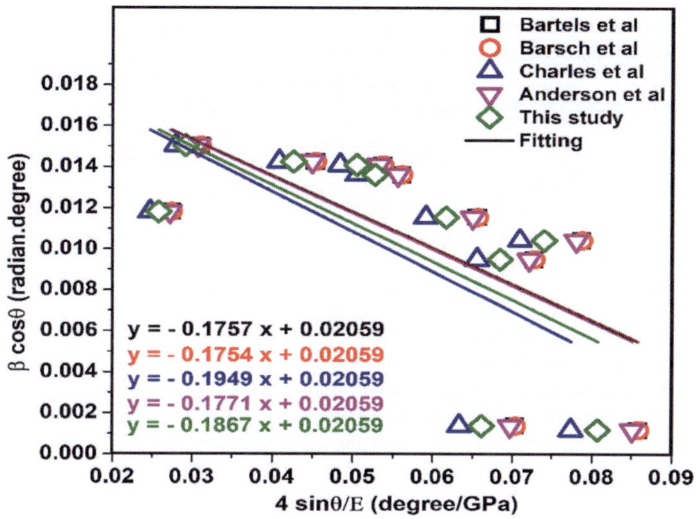

Fig. 3.9 Modified (W–H) USDM plot of NaCl

Table 3.6 Values derived of mechanical properties related to the NaCl

Mechanical properties					
Study	σ (GPa)	ε	μ [a] (GPa)	υ [b]	B (GPa) [c]
Expt. by (Bartels et al.)	−0.1757	−0.00523	14.91	0.24	24.71
Expt. by (Barsch et al.)	−0.1754	−0.00523	14.90	0.24	24.73
Expt. by (Charles et al.)	−0.1949	−0.00523	16.10	0.23	25.13
Theory. by (Anderson et al.)	−0.1771	−0.00523	14.93	0.25	25.30
This study	−0.1867	−0.00523	15.60	0.23	24.54

References

1. M. Rabiei, A. Palevicius, A. Dashti, S. Nasiri, A. Monshi, A. Vilkauskas, G. Janusas, Measurement modulus of elasticity related to the atomic density of planes in unit cell of crystal lattices. Materials. Basel. MDPI **13**(19), 1–17, art. no. 4380 (2020). ISSN 1996-1944. https://doi.org/10.3390/ma13194380
2. P. Ravindran, L. Fast, P.A. Korzhavyi, B. Johansson, Density functional theory for calculation of elastic properties of orthorhombic crystals: application to **TiSi2**. Appl. Phys. **84**, 4891 (1998). https://doi.org/10.1063/1.368733
3. M.E. Fine, L.D. Brown, H.L. Marcus, Elastic constants versus melting temperature in metals. Scr. Metall. **18**, 951–956 (1984). https://doi.org/10.1016/0036-9748(84)90267-9
4. M. Nakamura, Elastic constants some transition metal-disilicide-single crystals. Metall. and Mater. Trans. A. **25**, 331–340 (1994)
5. F. Brich, Finite elastic strain of cubic crystals. Phys. Rev **71**, 809 (1947). https://doi.org/10.1103/PhysRev.71.809
6. G. Leibfried, W. Ludwig, Theory of anharmonic effects in crystals. Solid State Phys. **12**, 275–444 (1961). https://doi.org/10.1016/S0081-1947(08)60656-6

7. G.R. Barsch, Z.P. Chang, Adiabatic, isothermal, and intermediate pressure derivatives of the elastic constants for cubic symmetry. II. Numerical results for 25 materials. Basic Solid States Phys. **19**, 139–151 (1967). https://doi.org/10.1002/pssb.19670190116.
8. V.A. Lubarda, Apparent elastic constants of cubic crystals and their pressure derivatives. Int. J. Non-Linear Mech. **34**, 5–11 (1999). https://doi.org/10.1016/S0020-7462(97)00068-1
9. F. Yu, Y. Liu, First-principles calculations of structural, mechanical, and electronic properties of the B2-phase NiTi shape-memory alloy under high pressure. Computation **7**, 57 (2019). https://doi.org/10.3390/computation7040057
10. D. Holec, M. Friák, J. Neugebauer, P.H. Mayrhofer, Trends in the elastic response of binary early transition metal nitrides. Phys. Rev. B **85**, 064101 (2012). https://doi.org/10.1103/PhysRevB.85.064101
11. K.M. Knowles, P.R. Howie, The directional dependence of elastic stiffness and compliance shear coefficients and shear moduli in cubic materials. J. Elast. **120**, 87–108 (2015). https://doi.org/10.1007/s10659-014-9506-1
12. Q. Fan, A new method of calculating planar density: the position-duplication-number method. J. Appl. Cryst **49**, 1454–1458 (2016). https://doi.org/10.1107/S1600576716010827
13. Heryanto1, B. Abdullah1, D. Tahir, Mahdalia, Quantitative analysis of X-Ray diffraction spectra for determine structural properties and deformation energy of Al, Cu and Si. IOP Conf. Ser.: J. Phys.: Conf. Ser. **1317**, 012052 (2019). https://doi.org/10.1088/1742-6596/1317/1/012052
14. F. Badawi, P. Villain, Stress and elastic-constant analysis by X-ray diffraction in thin films. J. Appl. Cryst **36**, 869–879 (2003). https://doi.org/10.1107/S0021889803002486
15. R.C. Graham, X-ray diffraction and the identification and analysis of clay minerals. Soil Sci. **164**, 72–73 (1999)
16. T. Hanabusa, J. Fukura, H. Fujiwara, J. Fukura, X-ray stress measurement on the cementite phase in steels. Bull. JSME **12**, 931–939 (1969). https://doi.org/10.1299/jsme1958.12.931
17. Y.T. Prabhu, K.V. Rao, V.S.S. Kumar, B.S. Kumari, X-ray analysis by williamson-hall and size-strain plot methods of ZnO nanoparticles with fuel variation. World J. Nano Sci. Eng. **4**, 21–28 (2014). https://doi.org/10.4236/wjnse.2014.41004
18. M. Levy, Introduction to fundamentals of elastic constants. Exp. Methods Phys. Sci. **39**, 1–35 (2001). https://doi.org/10.1016/S1079-4042(01)80084-9
19. K. Rajabi, Sh., Hosseini-Hashemi, Application of the generalized Hooke's law for viscoelastic materials (GHVMs) in nanoscale mass sensing applications of viscoelastic nanoplates: a theoretical study. Eur. J. Mech. A. Solids **67**, 71–83 (2018). https://doi.org/10.1016/j.euromechsol.2017.08.015
20. J.P. Perdew, K. Burke, M. Ernzerhof, Generalized gradient approximation made simple. Phys. Rev. Lett. **77**, 3865 (1996). https://doi.org/10.1103/PhysRevLett.77.3865
21. J.P. Perdew, A. Ruzsinszky, G.I. Csonka, O.A. Vydrov, G.E. Scuseria, L.A. Constantin, X. Zhou, K. Burke, Restoring the density-gradient expansion for exchange in solids and surfaces. Phys. Rev. Lett. **102**, 039902 (2009). https://doi.org/10.1103/PhysRevLett.100.136406
22. L.D. Landau, J.S. Bell, M.J. Kearsley, L.P. Pitaevskii, E.M. Lifshitz, J.B. Sykes, Electrodynamics of continuous media **8** (1970) ISBN: 9781483293752
23. M. Shafiq, S. Arif, I. Ahmad, S. Jalali Asadabadi, M. Maqbool, H.A.R. Aliabad, Elastic and mechanical properties of lanthanide monoxides. J. Alloy. Compd. **618**, 292–298 (2015). https://doi.org/10.1016/j.jallcom.2014.08.171
24. Y. Li, Relations between elastic constants Cij and texture parameters for hexagonal materials. J. Appl. Phys. **67**, 2663 (1990). https://doi.org/10.1063/1.345479
25. D.A. Papaconstantopoulos, M.J. Mehl, Tight-binding method in electronic structure, in *Encyclopedia of Condensed Matter Physics*, pp. 194–206 (2005). https://doi.org/10.1016/B0-12-369401-9/00452-6
26. H.B. Huntington, The elastic constants of crystals. Solid State Phys. **7**, 213–351 (1958). https://doi.org/10.1016/S0081-1947(08)60553-6
27. J.M. Zhang, Y. Zhang, K.W. Xu, V. Ji, Young's modulus surface and Poisson's ratio curve for cubic metals. J. Phys. Chem. Solids **68**, 503–510 (2007). https://doi.org/10.1016/j.jpcs.2007.01.025

28. J.M. Zhang, Y. Zhang, K.W. Xu, V. Ji, Anisotropic elasticity in hexagonal crystals. Thin Solid Films **515**, 7020–7024 (2007). https://doi.org/10.1016/j.tsf.2007.01.045
29. M. Rabiei, A. Palevičius, A. Monshi, S. Nasiri, A. Vilkauskas, G. Janušas, Comparing methods for calculating nano crystal size of natural hydroxyapatite using X-Ray diffraction. Nanomater.-MDPI **10**, 1627 (2020). https://doi.org/10.3390/nano10091627
30. R. Hertzberg, Deformation and fracture mechanics of engineering materials (1976)
31. V. Berdichevsky, Energy of dislocation networks. Int. J. Eng. Sci. **103**, 35–44 (2016). https://doi.org/10.1016/j.ijengsci.2016.02.011
32. R. Abbaschian, L. Abbaschian, *Reed-Hill*, R.E. Physical Metallurgy Principles (2009). ISBN-13: 978-0-495-08254-5
33. J.F. Shackelford, *Introduction to Materials Science for Engineers, 8th Edition* (2015). ISBN-13: 9780133826654
34. W.D. Callister, D.G. Rethwisch, *Materials Science and Engineering 9E* (2010). ISBN: 978-1-118-32457-8
35. I.P. Jones, W.B. Hutchinson, Stress-state dependence of slip in Titanium-6Al-4V and other H.C.P. metals. Acta Metall. **29**, 951–968 (1981). https://doi.org/10.1016/0001-6160(81)90049-3
36. M.P. Miller, J.S. Park, P.R. Dawson, T.S. Han, Measuring and modeling distributions of stress state in deforming polycrystals. Acta Mater. **56**, 3927–3939 (2008). https://doi.org/10.1016/j.actamat.2008.04.062
37. T. Ungar, G. Ribarik, L. Balogh, A. Salem, S.L. Semiatin, G.B.M. Vaughan, Burgers vector population, dislocation types and dislocation densities in single grains extracted from a polycrystalline commercial-purity Ti specimen by X-ray line-profile analysis. Scripta Materialia **63**, 69–72 (2010). https://doi.org/10.1016/j.scriptamat.2010.03.016
38. E. Goettler, Dislocation structure and work-hardening of copper sinqle crystals with [100] axis orientation I. Dislocation arrangement and cell structure of crystals deformed in tension. Struct. Prop. Condens. Matter **28**, 1057–1076 (1973). https://doi.org/10.1080/14786437308220968
39. S. Elliott, *The Physics and Chemistry of Solids* (1998). ISBN: 978-0-471-98195-4
40. P.E. McHugh, Introduction to crystal plasticity theory, in *Mechanics of Microstructured Materials*, pp. 125–171 (2004)
41. N. Pandech, K. Sarasamak, S. Limpijumnong, Elastic properties of provskite ATiO3 (A=Be, Mg, Ca, Sr, and Ba) and PbBO3 (B = Ti, Zr, and Hf): first principles calculations. J. Appl. Phys. **117**, 174108 (2015). https://doi.org/10.1063/1.4919837
42. C. Suryanarayana, M.G. Norton, Practical aspects of x-ray diffraction, in *X-Ray Diffraction*, pp. 63–94 (1998). https://doi.org/10.1007/978-1-4899-0148-4_3
43. D. Nath, F. Singh, R. Das, X-ray diffraction analysis by Williamson-Hall, Halder-Wagner and size-strain plot methods of CdSe nanoparticles a comparative study. Mater. Chem. Phys. **239**, 122021 (2020). https://doi.org/10.1016/j.matchemphys.2019.122021
44. E. Deligoz, Y. Colakoglu, Y. Ciftci, Elastic, electronic, and lattice dynamical properties of CdS, CdSe, and CdTe. Physica B **373**, 124–130 (2006). https://doi.org/10.1016/j.physb.2005.11.099
45. N. Kishore, V. Nagarajan, R. Chandiramouli, Mechanical properties and band structure of CdSe and CdTe nanostructures at high pressure—a first-principles study. Process. Appl. Ceram. **13**, 124–131 (2019). https://doi.org/10.2298/PAC1902124K
46. T.B. Bateman, Elastic moduli of single-crystal zinc oxide. J. Appl. Phys. **33**, 3309 (1962). https://doi.org/10.1063/1.1931160
47. D. Bhatia, H. Sharma, R.S. Meena, V.R.A. Palkar, Novel ZnO piezoelectric microcantilever energy scavenger: fabrication and characterization. Sens. Bio-Sens. Res. **9**, 45–52 (2016). https://doi.org/10.1016/j.sbsr.2016.05.008
48. E. Emil, S. Gurmen, Estimation of yttrium oxide microstructural parameters using the Williamson-Hall analysis. Mater. Sci. Technol. **34**, 1549–1557 (2018). https://doi.org/10.1080/02670836.2018.1490857
49. O. Yeheskel, O. Tevet, Elastic moduli of transparent Yttria. Am. Ceram. Socity **82**, 136–144 (1999). https://doi.org/10.1111/j.1151-2916.1999.tb01733.x

50. H.A. Badehian, H. Salehi, M. Ghoohestani, First-principles study of elastic, structural, electronic, tehrmody-namical, and optical properties of Yttria (Y2O3) ceramic in cubic phase. Am. Ceram. Socity **96**, 1832–1840 (2013). https://doi.org/10.1111/jace.12259
51. A. Migliori, J. Sarrao, *Resonant Ultrasound Spectroscopy: Applications to Physics, Materials Measurements, and Nondestructive Evaluation* (1997). ISBN-13: 978-0471123606
52. H. Ledbetter, Elastic constants of body-centered-cubic titanium monocrystals. Appl. Phys. **95**, 4642 (2004). https://doi.org/10.1063/1.1688445
53. R.A. Bartels, D.E. Schuele, Pressure derivatives of the elastic constants of NaCl and KCl at 295°K and 195°K. J. Phys. Chem. Solids **26**, 537–549 (1965). https://doi.org/10.1016/0022-3697(65)90130-7
54. G.R. Barsch, Z.P. Chang, Adiabatic, isothermal, and intermediate pressure derivatives of the elastic constants for cubic symmetry. II. Numerical results for 25 materials. Phys. Stat. Sol **19**, 139–151 (1967). https://doi.org/10.1002/pssb.19670190116
55. H. Charles, Determination of the elastic constants of NaCl to 40 kilobars by brillouin scattering. Thesis of Master science, The Institute of Optics the University of Rochester Rochester, New York, p. 14627 (1975)
56. O.L. Anderson, K. Masuda, D.G.A. Isaak, New thermodynamic approach for high-pressure physics. Phys. Earth Planet. Inter. **91**, 3–16 (1995). https://doi.org/10.1016/0031-9201(95)03044-w
57. S.M. Abdou, H. Moharam, Characterization of table salt samples from different origins and ESR detection of the induced effects due to gamma irradiation. IOP Conf. Ser.: J. Phys.: Conf. Ser. **1253**, 012036 (2019). https://doi.org/10.1088/1742-6596/1253/1/012036
58. S.H. Jury, R.C. Ernst, The first approximate conditions for the formation of liquid crystals in solution. J. Phys. Colloid Chem. **53**(5), 609–624 (1949). https://doi.org/10.1021/j150470a002
59. H. Wang, P.L. Prendiville, P. McDonnell, V.W. Chang, An ultrasonic technique for the measurement of the elastic moduli of human cornea. J. Biomech. **29**, 1633–1636 (1996). https://doi.org/10.1016/S0021-9290(96)80017-0
60. D.E. Bray, R.K. Stanley, *Nondestructive Evaluation: A Tool in Design, Manufacturing and Service* p. 586 psl (1997)
61. R.S. Figliola, D.E. Beasley, *Theory and Design for Mechanical Measurements,Hoboken* (2011). ISBN: 978-1-119-47565-1
62. S. Adachi, *Handbook on Physical Properties of Semiconductors* (Springer, New York, 2004)
63. M. Bodke, U. Gawai, A. Patil, B. Dole, Estimation of accurate size, lattice strain using Williamson-Hall models, SSP and TEM of Al doped ZnO nanocrystals. J. Matériaux Tech. **106**, 602 (2018). https://doi.org/10.1051/mattech/2018055
64. A.N. Norris, Poisson's ratio in cubic materials. Proc. R. Soc **462**, 3385–3405 (2006). https://doi.org/10.1098/rspa.2006.1726
65. L.X.Z. Wua, H.F. Feng, S. Wu, High-pressure effect on elastic constants, stacking fault energy and correlation with dislocation properties in MgO and CaO. Eur. Phys. J. B. **85**, 226 (2012). https://doi.org/10.1140/epjb/e2012-30032-4
66. F.S. Khan, P.B. Allen, Temperature dependence of the elastic constants of p+ silicon. Phys. Status Solidi **128**, 31–38 (1985). https://doi.org/10.1002/pssb.2221280104
67. E. G¨uler, M. G¨uler, Elastic and mechanical properties of cubic diamond under pressure. Chin. J. Phys. **53**, 040807–040811 (2015). https://doi.org/10.6122/CJP.20141230

Chapter 4
Correlation Between Young's Modulus and Planar Density of Unit Cell, Super Cells (2 x 2 x 2), Symmetry Cells of Perovskite ($CaTiO_3$) Lattice

This chapter exemplifies the introduction of a novel method for determining the Young's modulus of crystallite structures, including $CaTiO_3$ (SC) [1]. Calcium titanate ($CaTiO_3$) was produced using the solvothermal process as another instance of a cubic lattice structure. Due to its dopant/doping mechanisms, calcium titanate-$CaTiO_3$ (a perovskite) has been applied in numerous industrial settings. Altering defective grain boundaries within the perovskite structure is critical for enhancing mechanical properties and stability, drawing significant attention to this structure. Without thoroughly understanding the perovskite's configuration and diffracted planes, dopant/doping mechanisms remain elusive. This study focused on the design and analysis of atom locations and diffracted planes. It explored the connection between Young's modulus and the planar density of the unit cell, super cells (2 × 2 × 2), and symmetry cells of nano $CaTiO_3$. The elastic constants, compliance, and Young's modulus were determined using the ultrasonic pulse-echo technique, with findings of C11 = 330.89 GPa, C12 = 93.03 GPa, C44 = 94.91 GPa, and E = 153.87 GPa, respectively. Young's modulus values derived from planar density were 162.62 GPa for the unit cell, 151.71 GPa for super cells (2 × 2 × 2), and 152.21 GPa for symmetry cells. The Young's modulus from planar density of symmetry cells aligned closely with the value obtained through ultrasonic pulse-echo.

Supplementary Information The online version contains supplementary material available at https://doi.org/10.1007/978-3-032-12583-5_4.

M. Rabiei et al., *Hydroxyapatite-Based Nanocomposites*, Springer Series in Chemical Physics 127, https://doi.org/10.1007/978-3-032-12583-5_4

4.1 Introduction

Perovskites have a general formula of ABO_3. In these structures, the A site cation is a typical lanthanide, alkaline or alkaline-earth metal with 12-fold oxygen coordination, and the B-site is any one of a variety of transition metal cations. Calcium titanate ($CaTiO_3$) was established in 1839 by a Russian mineralogist Perovski, and materials with the same type of $CaTiO_3$ were introduced as the perovskite structure. $CaTiO_3$ has ionic bonds, as well as the ionic radii of Ca^{2+}, O^{2-} and Ti^{4+} are 1 Å, 1.40 Å and 0.6 Å, respectively [2]. In recent years, researchers have concentrated on innovating perovskites and exploring their mechanical attributes to achieve a high yield. Additionally, CaTiO3 is a well-known member of the ferroelectric perovskite family, extensively used as a dopant in electronic applications due to its dielectric properties and adaptability in structural shifts [2, 3]. The modulus of elasticity (E) or Young's modulus is defined as the ratio of stress to strain when the material is within its elastic range [4]. Elasticity constants are determined through the deformation of crystal lattices under applied force. These include Young's modulus, shear modulus, and volumetric modulus, inherently associated with a material's elastic properties and its resistance to deformation under load. The elastic behavior of materials is characterized by models such as Cauchy, hypo-elastic, and hyper-elastic. A hyper-elastic model pertains to ideally elastic materials reacting to stress from a strain energy density function, whereas a hypo-elastic model's governing equation, apart from its linear state, is unaffected by finite strain magnitude [5]. The elastic properties are closely linked to the crystal structure, the intrinsic bonding nature between atoms, and the materials' anisotropic characteristics [6, 7]; thus, these elastic constants can be determined from crystal lattice computations [8]. Numerous studies have examined the correlation between elastic constants and lattice structure planes/directions, as noted in references [9–11]. The ultrasonic pulse-echo technique is one of the most precise methods for determining elastic stiffness constants and Young's modulus, relying on the velocity of long-wavelength acoustic waves [12]. In crystal structures, an indexing scheme describes points, directions, and planes, with planar density defined as the atom count per unit area on a specific crystallographic plane with a designated index [13]. Since X-rays were discovered in the late 19th century, they have been commonly employed for material characterization [14], allowing determination of atomic-scale structures across various material states [15]. X-ray diffraction uniquely specifies both the mechanical and microstructural characteristics of each diffracted plane, which serve as a strain gauge to measure Young's modulus in specific diffraction vector planes/directions [16]. In equipment manufacturing and design, reliable, non-destructive methods for assessing material mechanical properties are crucial. Traditional mechanical tests like tensile and impact tests are destructive, while ultrasonic approaches demand significant time and expertise, and theoretical methods such as density functional theory (DFT) calculations require substantial time and might need experimental validation. Our proposed method relies solely on X-ray diffraction (XRD) analysis and planar density calculation, making it highly relevant for industrial use. This study examines how cell size affects the precision of Young's modulus calculations, focusing on atom locations and diffracted planes in unit cells, 2 × 2 × 2 supercells, and symmetry cells of CaTiO3. A supercell has the same crystal structure as a unit cell but encompasses a larger volume. Supercells are created by proportionally extending a unit cell according to lattice vectors. For instance, a 2 × 2 × 2 supercell represents twice the unit cell's length in each direction, while an 8 × 8 × 8 supercell extends it eightfold. Results obtained from symmetry cells aligned closely with those from ultrasonic techniques. Consequently, we propose this novel XRD-based approach for determining reliable Young's modulus in single crystal or polycrystalline CaTiO3.

4.2 Experimental

4.2.1 Materials

In this study, for synthesis $CaTiO_3$, titanium (IV) butoxide, calcium chloride dehydrate, sodium hydroxide and ethanol reagents were purchased from Sigma Aldrich (Taufkirchen, Germany) and deionized water as the solvent for dispersions was prepared.

4.2.2 Instrumentation

In this research, a Bruker D8 Advance X-Ray diffractometer (Kaunas, Lithuania) with $Cu_{K\alpha}$ radiation was used. The powder X-ray diffraction was taken at 40 kV and 40 mA, and it was registered at a scanning rate of 2.5 degrees/minute and a step size of 0.02 degrees. The XRD peaks were interpreted by High Score X'Pert software (4.9.0) analysis to get the output ASC type files. The pulse-echo method was employed to measure the velocity of sound for both transverse and longitudinal ultrasonic waves. The ultrasonic measurements utilized a Panametrics Co. (Waltham, MA, USA) pulser receiver and an Iwatsu (Tokyo, Japan) oscilloscope, which operated at 100 MHz. The powder was shaped using CD04-Z mechanical presses and a CIP (CP 360). The specific surface area of the specimen was examined by measuring nitrogen (N\textsubscript{2}) gas desorption isotherms with a Brunauer-Emmett-Teller (BET) apparatus from Gemini V analyzer, micrometrics GmbH (Tehran, Iran). Transmission electron microscopy (TEM) was performed using a CM 10-Philips (Tehran, Iran) microscope, with an acceleration voltage range of 50–80 kV.

4.2.3 *Methods*

4.2.3.1 Synthesis of Nano-Powder $CaTiO_3$

The pulse-echo technique was utilized to assess the sound velocity in both transverse and longitudinal ultrasonic waves. For these ultrasonic measurements, a pulser receiver from Panametrics Co. (Waltham, MA, USA) and a 100 MHz oscilloscope from Iwatsu (Tokyo, Japan) were employed. The powder was formed using CD04-Z mechanical presses along with a CIP (CP 360). The sample's specific surface area was analyzed by determining nitrogen (N\textsubscript{2}) gas desorption isotherms using a Brunauer–Emmett–Teller (BET) device from the Gemini V analyzer by micrometrics GmbH (Tehran, Iran). Transmission electron microscopy (TEM) was conducted with a CM 10-Philips microscope (Tehran, Iran) at an acceleration voltage of 50–80 kV.

4.2.3.2 X-Ray Diffraction of $CaTiO_3$ and Planar Density Calculations

X-ray diffraction on crystalline $CaTiO_3$ was carried out alongside calculations of planar density for each diffracted plane. In our research, the atomic density of a plane was defined as the planar density, which is calculated by dividing the area occupied by atoms centered on the plane by the total area of the plane. This parameter significantly influences the mechanical characteristics of each plane. Planar density is dimensionless and its value is always less than 1 for each cell. Additionally, planar density values depend on the atomic arrangements and environments within the planes. The software Crystal Maker, Version 10.2.2 was utilized to determine atomic areas. Initially, the 3D geometric configurations of the crystal structures were developed, and then the atomic area was derived from the intersecting regions of each diffracted plane with the atoms present in the plane. When an atom with diameter D was involved completely, the atomic area will be $A = \pi(\frac{D}{2})^2$; otherwise, it will be a percentage of this amount.

4.2.3.3 Ultrasonic Pulse-Echo Technique of $CaTiO_3$

Ultrasonic waves, a form of elastic waves moving through a medium at high frequencies, are used to determine the Young's modulus of samples. By understanding the parameters of ultrasonic waves, more precise mechanical properties can be measured [19]. Recently, ultrasonic methods have become a focus of research into mechanical properties. Essentially, longitudinal and transverse waves pass through nano- or microstructures at varying speeds, with each velocity reflecting specific material properties.

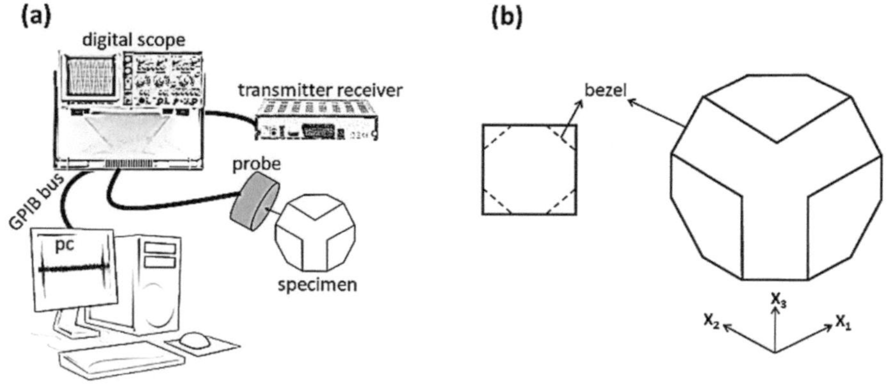

Fig. 4.1 Schematic of **a** ultrasonic pulse instrument and **b** a sketch of prepared $CaTiO_3$ sample

For ultrasonic measurements based on the Christoffel procedure, the first cubic specimen of $CaTiO_3$ was prepared by cold isostatic press. The schematics of ultrasonic measurement are depicted in Fig. 4.1a. The main part of the ultrasonic system is the pulser-receiver, which creates an electric pulse and stimulates the probe. Furthermore, the produced pulses enter the specimen, and after a sweep, they can be received via a probe. In this measurement, some drops of water were utilized to prevent the depreciation of waves in the air, and the effect of hand pressure on the probe was decreased [20].

At any position in the sample, a local coordinate is adjusted, such as X_1, the radial coordinate; X_2, the circumferential coordinate; and X_3, the axial coordinate. $V_{i/j}$ denotes the velocity of an ultrasound wave propagating in the X_i direction with particle displacements in the X_j direction. $V_{i/j}$ with the same i and j is longitudinal, and with $i \neq j$ is related to the transverse waves. For the measurement of quasi-longitudinal or quasi-transverse velocity ($V_{ij/ij}$), the specimen should be cut (bezel) on the edges of the surfaces perpendicular to the X directions. A sketch of the sample is represented in Fig. 4.1b.

4.3 Results

4.3.1 X-Ray Diffraction of $CaTiO_3$ and Planar Density Calculations

The XRD pattern of $CaTiO_3$ is presented in Fig. 4.2. The characteristic peaks of $CaTiO_3$ correspond to the report in Ref [21]. The crystal structure of $CaTiO_3$ is cubic, the atomic positions of Ti are at (000), Ca at $(\frac{1}{2}, \frac{1}{2}, \frac{1}{2})$ and O at $(\frac{1}{2}, 0,0)$, $(0, \frac{1}{2}, 0)$, $(0,0, \frac{1}{2})$. According to X-ray powder diffraction results, the lattice parameter

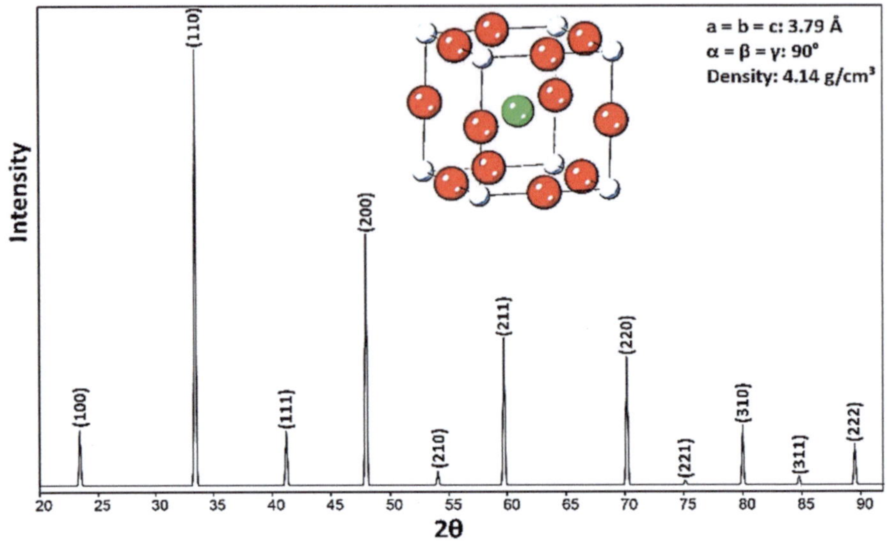

Fig. 4.2 X-ray diffraction of $CaTiO_3$ (powder sample)

is 3.79 ± 0.02 Å, which is in good corresponds with the amount recorded in the Ref [22]. In addition, crystallographic parameters (Table S8.2.1) of $CaTiO_3$ and analyzed data by X'Pert [23] nasiri are recorded as the cell volume = 54.44 ±0.01 Å ^{3} and crystal density = 4.14 ± 0.01 g/cm^3, and the space group is Pm-3 m. In addition, the crystal size of $CaTiO_3$ was calculated by the Monshi–Scherrer equation (Fig. S8.2.2) [24] and BET analysis. The crystal size values were registered at ~59.10 and 63.02 nm, respectively. The Monshi–Scherrer method is described in Sect. 2 of the supporting information. Furthermore, a TEM image of $CaTiO_3$ is shown in Fig. S8.2.3. According to the images shown in Fig. S8.2.3, the size of $CaTiO_3$ particles basically corresponds to the crystallite size, and it is clear that particles of powder have nanoscale and size can be reported almost ± 50 nm.

For the evaluation of cells as the results, the comprehensive calculations of the planar density of diffracted planes in the unit cell, super cells (2 × 2 × 2) and super cells (8 × 8 × 8) of $CaTiO_3$ lattice are presented in Figs. S8.2.4–S8.2.6 respectively. In addition, the locations of atoms, geometry of planes and calculations of planar density of (211) super cell (4 × 4 × 4), (211) super cell (8 × 8 × 8), (221) super cell (4 × 4 × 4), (221) super cell (8 × 8 × 8), (311) super cell (3 × 3 × 3), (311) super cell (4 × 4 × 4), (311) super cell (8 × 8 × 8), (222) super cell (3 × 3 × 3) and (222) super cell (8 × 8 × 8) are depicted briefly in Figs. 4.3–4.6 respectively. Furthermore, the completed calculations with their figures are shown in Figs. S8.2.7–S8.2.10.

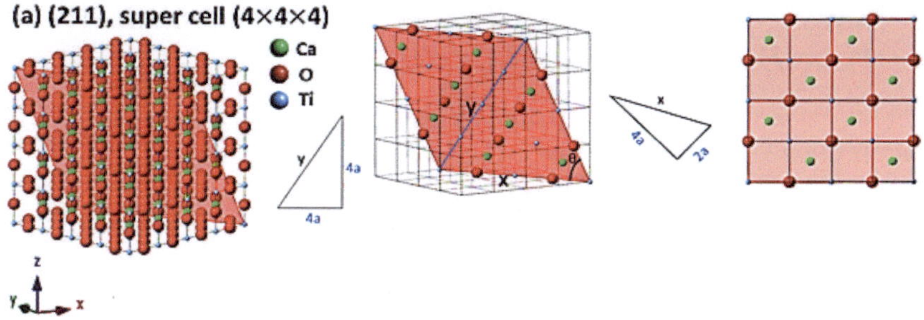

number of atoms in the plane (211) × area of each atom in the plane (211) =

$$\left[\left(\left(2\times\frac{78.52}{360}+2\times\frac{101.48}{360}+4\times\frac{1}{2}+5\right)\times\pi\left(r_{Ti^{4+}}\right)^2\right)+\left(\left(8\times\frac{1}{2}+4\right)\times\pi\left(r_{O^{2-}}\right)^2\right)\right]+\left[\left(8\times\pi\left(r_{Ca+}\right)^2\right)\right]=$$

$$\left[\left(8\times\pi\,(0.60)^2\right)+\left(8\times\pi\,(1.40)^2\right)+\left(8\times\pi\,(1)^2\right)\right]=83.44$$

$$\text{planer density}=\frac{\text{number of atoms in the plane (211)}\times\text{area of each atom in the plane (211)}}{\text{area of the plane (211)}}=\frac{83.44}{334.1}=0.25$$

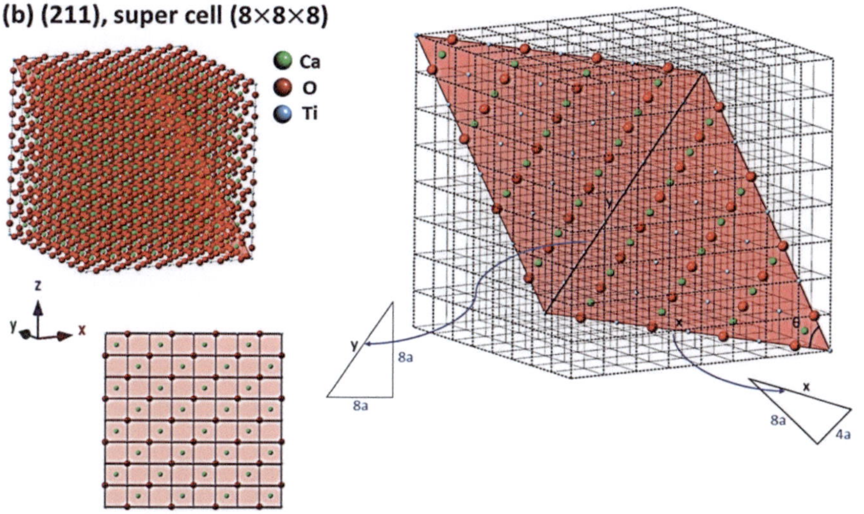

number of atoms in the plane (211) × area of each atom in the plane (211) =

$$\left[\left(\left(2\times\frac{78.52}{360}+2\times\frac{101.48}{360}+12\times\frac{1}{2}+25\right)\times\pi\left(r_{Ti^{4+}}\right)^2\right)+\left(\left(16\times\frac{1}{2}+24\right)\times\pi\left(r_{O^{2-}}\right)^2\right)\right]+\left[\left(32\times\pi\left(r_{Ca+}\right)^2\right)\right]=$$

$$\left[\left(32\times\pi\,(0.60)^2\right)+\left(32\times\pi\,(1.40)^2\right)+\left(32\times\pi\,(1)^2\right)\right]=333.76$$

$$\text{planer density}=\frac{\text{number of atoms in the plane (211)}\times\text{area of each atom in the plane (211)}}{\text{area of the plane (211)}}=\frac{333.76}{1335.24}=0.25$$

Fig. 4.3 Geometry of planes and calculations of planar density of **a** (211) super cell (4 × 4 × 4) and **b** (211) super cell (8 × 8 × 8) (which shows and emphasizes the symmetry of (8 × 8 × 8) super cells)

4.3.2 *Investigation of Results Obtained from Ultrasonic Pulse-Echo Technique of $CaTiO_3$*

Taking into account the Christoffel equation, the connection between ultrasonic phase velocity and the stiffness matrix is given as follows:

$$\left(C_{ijkl}l_jl_l - \rho V^2\delta_{ik}\right)\alpha_k = 0$$

where V is the ultrasonic phase velocity, C_{ijkl} is the general stiffness matrix, ρ is the material density, l is the orientation of propagation, α_k is the polarization direction and δ_{ik} is the Kronecker delta (note that i,j, k, I = 1 to 3). For the extraction and calculation of elastic constants from ultrasonic measurements based on the Christoffel equation, with the propagation in X_1, X_2 and X_3 directions, all of the diagonal elements of the stiffness matrix are obtained. For the determination of whole constants, we cut the specimen on the edges of the surfaces perpendicular to principal directions (bezel) and the velocity was measured from the propagation of ultrasound wave normal to these planes.

Based on Eqs. (4.1)–(4.5) [25, 26] and the measured velocity according to the Table 4.1, stiffness constants values were obtained. C_{11} is in the agreement with longitudinal distortion and longitudinal compression/tension, so C_{11} can be described as the hardness. Moreover, the transverse distortion is connected to the C_{12}, and C_{12} is obtained from the transverse expansion correlated to the Poisson's ratio. C_{44} is based on the shear modulus, as well as C_{12} is in the settlement with C_{11} and C_{12} [25].

$$C_{11} = \rho V^2_{1/1} \tag{4.1}$$

$$C_{22} = \rho V^2_{2/2} \tag{4.2}$$

$$C_{66} = \rho V^2_{1/2} = \rho V^2_{2/1} \tag{4.3}$$

$$C_{12} = \sqrt{\left(C_{11} + C_{66} - 2\rho V^2_{12/12}\right)\left(C_{22} + C_{66} - 2\rho V^2_{12/12}\right)} - C_{66} \tag{4.4}$$

$$C_{44} = \rho V^2_{2/3} = \rho V^2_{3/2} \tag{4.5}$$

After substitution and calculation, C_{11}, C_{12} and C_{44} were registered at 330.89, 93.03 and 94.91 GPa respectively. These values of $CaTiO_3$ were in good agreement with the values submitted in the [27–29]. Moreover, with the ultrasonic technique, longitudinal and transverse waves can be utilized for determining Young's modulus quantity [30, 31]. The longitudinal and transverse waves of $CaTiO_3$ sample are shown in Fig. 4.7. In this method, by measuring the waves velocity and density of specimen,

Table 4.1 The values of longitudinal and transverse velocity of the sample

Longitudinal velocity (m/s)	Transverse velocity (m/s)	Quasi longitudinal or quasi transverse (m/s)
$V_{1/1} = 9261.85$	$V_{2/3} = 4960.5$	$V_{12/12} = 4976.63$
$V_{2/2} = 8013.51$	$V_{1/2} = 4283.65$	

the determination of Young's modulus quantity was carried out (Eq. (4.6)).

$$E = \frac{4\rho\left(\frac{L}{t_s}\right)^2 (3t_s^2 - 4t_l^2)}{t_s^2 - t_l^2} \tag{4.6}$$

where, t_s and t_l are differences between two echo in longitudinal and transverse waves, respectively [32, 33]. According to the results shown in Fig. 4.7, t_s and t_l values are calculated as 5.75 and 3.01 μs, respectively. In addition, the density of the specimen is recorded as 3857.30 $\frac{Kg}{m^3}$, and the length of the specimen after powder pressing reached 11.21 mm. After calculation, Young's modulus value of $CaTiO_3$ was 153.87 GPa. This value corresponds with the value reported by Ramajo et al. [34].

4.3.3 *Calculations: Relationship between Elastic Stiffness-Compliance Constants, Young's Modulus and Planar Density Extracted through the Unit Cell, Super Cells (2 × 2 × 2) and Symmetry Cells of $CaTiO_3$ Lattice*

Three elastic constants of $CaTiO_3$ were calculated via the ultrasonic technique. For the cubic $CaTiO_3$ system, the relationship between stiffness (C_{ij}) and compliance constant (S_{ij}) are provided in Eqs. (4.7)–(4.9) [26, 35]. The values resulted via Eqs. (4.7)–(4.9) are 0.0034, −0.0007 and 0.0105 GPa for S_{11}, S_{12} and S_{44}, respectively. Furthermore, Young's modulus of each diffracted plane of $CaTiO_3$ can be written as Eq. (4.10) [36].

$$S_{11} = \frac{C_{11} + C_{12}}{(C_{11} - C_{12})(C_{11} + 2C_{12})} \tag{4.7}$$

$$S_{12} = \frac{-C_{12}}{(C_{11} - C_{12})(C_{11} + 2C_{12})} \tag{4.8}$$

$$S_{44} = \frac{1}{C_{44}} \tag{4.9}$$

Table 4.2 Planar density and Young's modulus values of the unit cell, super cells (2 × 2 × 2) and symmetry cells of $CaTiO_3$

Index	Planar density of unit cell	Planar density of super cell (2 × 2 × 2)	Planar density of symmetry cells	Planar density of super Cell (8 × 8 × 8)	Young's modulus (GPa)
(100)	0.93	0.93	0.93 in (2 × 2 × 2)	0.93	290.059
(110)	0.51	0.51	0.51 in (2 × 2 × 2)	0.51	221.652
(111)	0.04	0.04	0.04 in (2 × 2 × 2)	0.04	179.354
(200)	0.64	0.64	0.64 in (2 × 2 × 2)	0.64	290.059
(210)	0.41	0.41	0.41 in (2 × 2 × 2)	0.41	194.176
(211)	0.16	0.25	0.25 in (2 × 2 × 2)	0.25	150.612
(220)	0.6	0.6	0.6 in (2 × 2 × 2)	0.6	129.810
(221)	0.46	0.29	0.31 in (4 × 4 × 4)	0.31	109.622
(310)	0.24	0.24	0.23 in (4 × 4 × 4)	0.23	186.471
(311)	0.04	0.03	0.02 in (3 × 3 × 3)	0.02	140.386
(222)	0.99	0.88	0.88 in (3 × 3 × 3)	0.88	83.615

$$\frac{1}{E_{hkl}} = S_{11} - 2\left[(S_{11} - S_{12}) - \frac{1}{2}S_{44}\right] \cdot \left[\frac{h^2k^2 + k^2l^2 + l^2h^2}{(h^2 + k^2 + l^2)}\right] \tag{4.10}$$

(The planar density and Young's modulus values related to the each diffracted plane of the unit, super (2 × 2 × 2), symmetry and super (8 × 8 × 8) cells of $CaTiO_3$ lattice are tabulated in Table 4.2.

4.4 Discussion

Table 4.2 and Figs. 4.3, 4.4, 4.5 and 4.6 indicate that the optimum matrix is achieved with the expanded cells, identified here as symmetry cells. This optimum matrix represents the smallest extension of a unit cell plane to a super cell, beyond which the plane's density remains unchanged. For instance, the symmetry cell or optimum matrix for the (311) plane is (3 × 3 × 3), implying that both (4 × 4 × 4) and (8 × 8 × 8) matrices yield consistent planar density values (see Fig. 4.5a–c). Actual planar density values are derived from this symmetry cell, as extending a plane to infinity once its symmetry is achieved maintains consistent density. Identifying the symmetry cell requires knowledge of parameters such as crystal lattice arrangements, atom locations within planes, and plane indices. Thus, determining Young's modulus values based on $CaTiO_3$'s planar density necessitates identifying these symmetry cells. Notably, real or symmetrical values correlate with the (8 × 8 × 8) super cell, as this size provides complete lattice correspondence in all directions; hence, real planar density calculations should utilize the (8 × 8 × 8) super cell. To verify this, Figs. 4.3b, 4.4b, 4.5c, and 4.6b detail the real planar density and geometry of atoms and planes like (211), (221), (311), and (222) in these larger super cells, respectively. While determining the precise configuration of planes and geometries is complex, having this information improves the accuracy of results obtained from Young's modulus values. The primary hypothesis is that increased planar density demands greater forces for atomic movement via dislocation mechanisms. Dislocations, which are deviations in the lattice caused by adding or removing a plane of atoms from an otherwise perfect crystal, result in a loss of acoustic energy, thereby impacting the wavelength and duration of ultrasonic waves [37].

The force (W), which is needed for the movement of atoms in each plane, is obtained from Eq. (4.11) [38].

$$W = \frac{\mathrm{E}}{2(1+\nu)} b^2 1 \tag{4.11}$$

In Eq. (4.11), E is Young's modulus, b is Burgers vector, l is dislocation length and ν is Poisson's ratio. The higher value of force is in accordance with the modulus of elasticity (Young's modulus), which would be higher.

To compare Young's modulus values of $CaTiO_3$ in a unit cell, super cells (2 × 2 × 2) and symmetry cells, the fitting of Young's modulus values extracted by each diffracted plane versus planar density values is presented in Fig. 4.8. According to the results (shown in the Fig. 4.8) and the straight fitting line, Young's modulus values of unit cell, super cells (2 × 2 × 2) and symmetry cells were calculated as 162.62 ± 0.4 GPa, 151.71 ± 0.4 GPa and 152.21 ± 0.4 GPa, respectively. As expected, the Young's modulus value of symmetry cells of $CaTiO_3$ (152.21 ± 0.4 GPa) is in good agreement with experimental Young's modulus value extracted via ultrasonic-echo technique (153.87 ± 0.2 GPa). Moreover, Young's modulus value of unit cell (162.62 ± 0.4 GPa) has a greater difference with experimental Young's modulus value, and as a result, the unit cell of $CaTiO_3$ cannot be represented as whole cells. This is because in a unit cell of $CaTiO_3$, crystalline defects are not considered and is especially controlling of deformation, and displacement of atoms in the planes is related to the dislocation networks [39]. Further, a unit cell of $CaTiO_3$ is not involved in imperfections (such as dislocations, Frenkel defect and Schottky defect) with respect to the super cell [40]; therefore, the slope line value of the unit cell is reported (37.23) to be less than the slope line value of super cells (2 × 2 × 2) (63.67) and symmetry cells (62.41). Consequently, the effect of imperfections in expanded cells (super cells) is very impressive, so the unit cell of $CaTiO_3$ is considered as the

(a) (221), super cell (4×4×4)

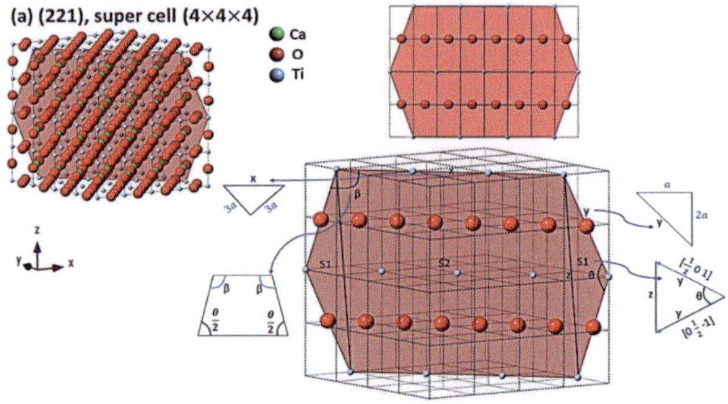

number of atoms in the plane (221) × area of each atom in the plane (221) =

$$\left[\left(\left(2\times\frac{143.13}{360}+4\times\frac{108.44}{360}+4\times\frac{1}{2}+3\right)\times\pi\left(r_{Ti^{4+}}\right)^2\right)+\left(\left(12+4\times\frac{1}{2}\right)\times\pi\left(r_{O^{2-}}\right)^2\right)\right]=\left[\left(7\times\pi\,(0.60)^2\right)+\left(14\times\pi\,(1.40)^2\right)\right]=94.12$$

$$\text{planer density}=\frac{\text{number of atoms in the plane (221)}\times\text{area of each atom in the plane (221)}}{\text{area of the plane (221)}}=\frac{94.12}{301.71}=0.31$$

(b) (221), super cell (8×8×8)

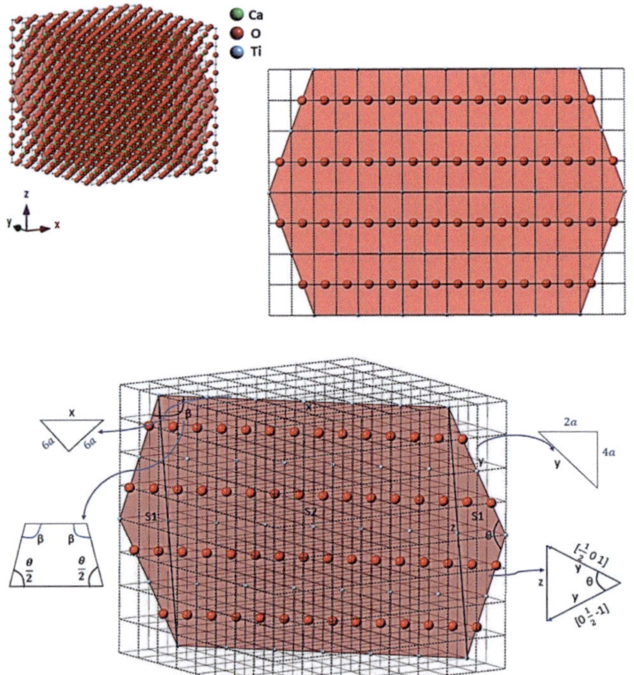

number of atoms in the plane (221) × area of each atom in the plane (221) =

$$\left[\left(\left(2\times\frac{143.13}{360}+4\times\frac{108.44}{360}+14\times\frac{1}{2}+19\right)\times\pi\left(r_{Ti^{4+}}\right)^2\right)+\left(\left(52+8\times\frac{1}{2}\right)\times\pi\left(r_{O^{2-}}\right)^2\right)\right]=\left[\left(28\times\pi\,(0.60)^2\right)+\left(56\times\pi\,(1.40)^2\right)\right]=376.49$$

$$\text{planer density}=\frac{\text{number of atoms in the plane (221)}\times\text{area of each atom in the plane (221)}}{\text{area of the plane (221)}}=\frac{376.49}{1206.65}=0.31$$

Fig. 4.4 Geometry of planes and calculations of planar density of **a** (221) super cell (4 × 4 × 4) and **b** (221) super cell (8 × 8 × 8) (which shows and emphasizes the symmetry of (8 × 8 × 8) super cells)

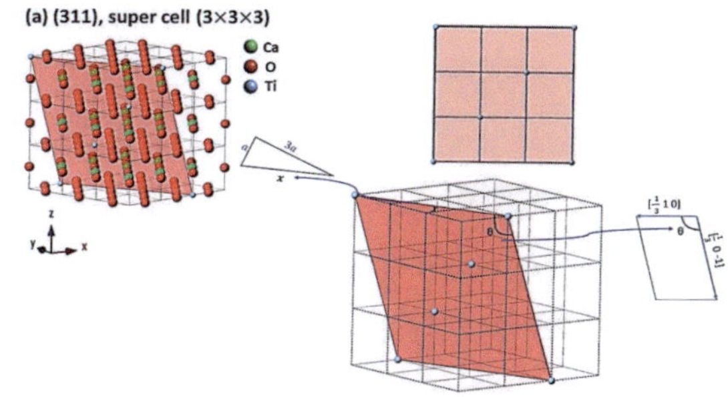

number of atoms in the plane (311) × area of each atom in the plane (311) =

$$\left[\left(2\times\frac{95.74}{360}+2\times\frac{84.26}{360}+2\right)\times\pi\left(r_{Ti^{4+}}\right)^2\right]=(3\times\pi\times(0.6)^2)=3.39$$

$$\text{planer density}=\frac{\text{number of atoms in the plane (311)}\times\text{area of each atom in the plane (311)}}{\text{area of the plane (311)}}=\frac{3.39}{143.04}=0.02$$

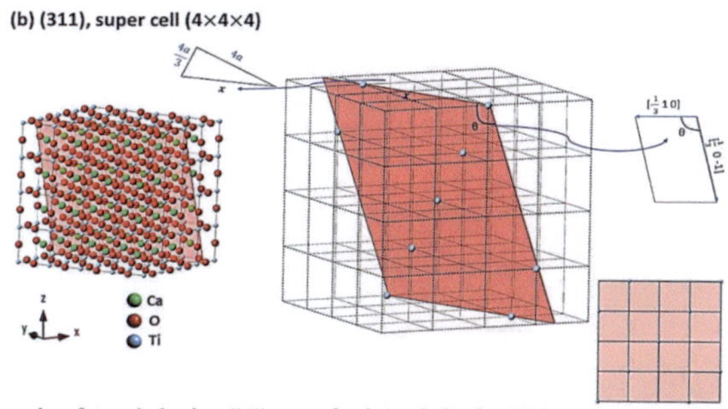

number of atoms in the plane (311) × area of each atom in the plane (311) =

$$\left[\left(2\times\frac{95.74}{360}+4\times\frac{1}{2}+3\right)\times\pi\left(r_{Ti^{4+}}\right)^2\right]=(5.53\times\pi\times(0.6)^2)=6.25$$

$$\text{planer density}=\frac{\text{number of atoms in the plane (311)}\times\text{area of each atom in the plane (311)}}{\text{area of the plane (311)}}=\frac{6.25}{254.40}=0.02$$

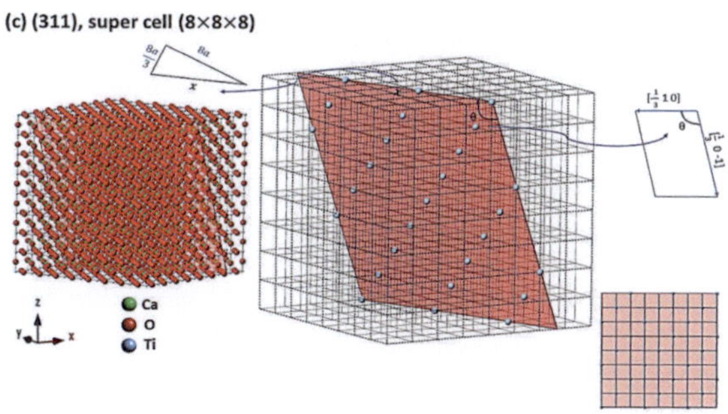

number of atoms in the plane (311) × area of each atom in the plane (311) =

$$\left[\left(2\times\frac{95.74}{360}+8\times\frac{1}{2}+17\right)\times\pi\left(r_{Ti^{4+}}\right)^2\right]=(21.53\times\pi\times(0.6)^2)=24.35$$

$$\text{planer density}=\frac{\text{number of atoms in the plane (311)}\times\text{area of each atom in the plane (311)}}{\text{area of the plane (311)}}=\frac{24.35}{1016.32}=0.02$$

Fig. 4.5 The concept of a symmetry cell; geometry of planes and calculations of planar density of **a** (311) super cell (3 × 3 × 3), **b** (311) super cell (4 × 4 × 4) and (**c**) (311) super cell (8 × 8 × 8)

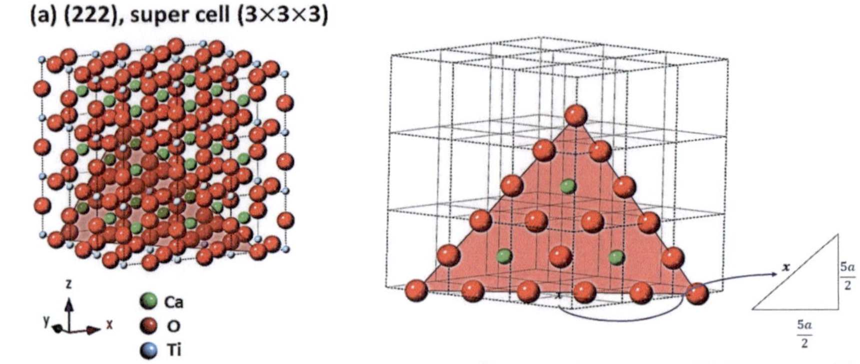

number of atoms in the plane (222) × area of each atom in the plane (222) $= \left[\left((3 \times \frac{1}{6} + 12 \times \frac{1}{2} + 3) \times \pi \left(r_{O^{2-}}\right)^2\right)\right] + \left[\left(3 \times \pi \times \left(r_{Ca^{+}}\right)^2\right)\right] =$

$[(9.5 \times \pi (1.40)^2)] + [(3 \times \pi (1)^2)] = 67.92$

$$\text{planer density} = \frac{\text{number of atoms in the plane (222)} \times \text{area of each atom in the plane (222)}}{\text{area of the plane (222)}} = \frac{67.92}{77.28} = 0.88$$

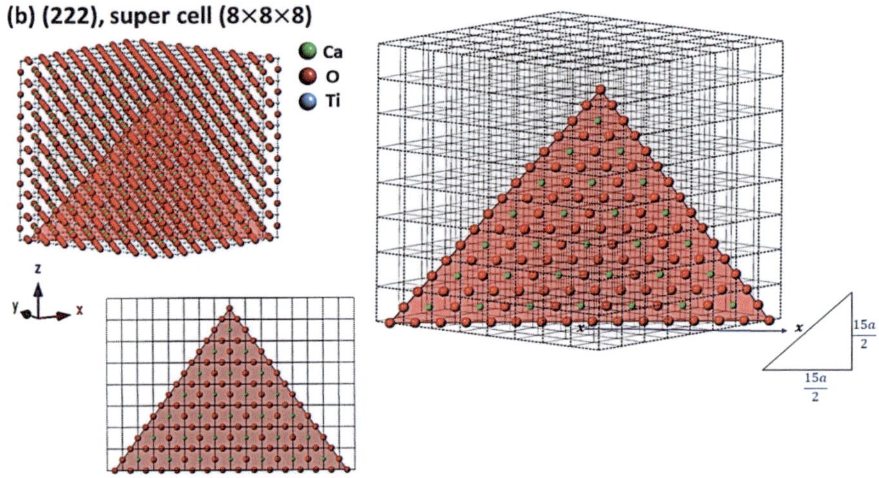

number of atoms in the plane (222) × area of each atom in the plane (222) $= \left[\left((3 \times \frac{1}{6} + 42 \times \frac{1}{2} + 63) \times \pi \left(r_{O^{2-}}\right)^2\right)\right] + \left[\left(28 \times \pi \times \left(r_{Ca^{+}}\right)^2\right)\right] =$

$[(84.5 \times \pi (1.40)^2)] + [(28 \times \pi (1)^2)] = 608.28$

$$\text{planer density} = \frac{\text{number of atoms in the plane (222)} \times \text{area of each atom in the plane (222)}}{\text{area of the plane (222)}} = \frac{608.28}{695.92} = 0.88$$

Fig. 4.6 Geometry of planes and calculations of planar density of **a** (222) super cell (3 × 3 × 3) and **b** (222) super cell (8 × 8 × 8)

ideal lattice, while symmetry cells such as (8 × 8 × 8) of $CaTiO_3$ are real lattices [41]; this is consistent with the experimental Young's modulus. It is clear that each imperfection will be caused by a decreasing Young's modulus [42], and in Fig. 4.8, this matter is confirmed when the Young's modulus value (intercept) in the unit cell of $CaTiO_3$ is higher than in super cells (2 × 2 × 2) and symmetry cells. Apparently,

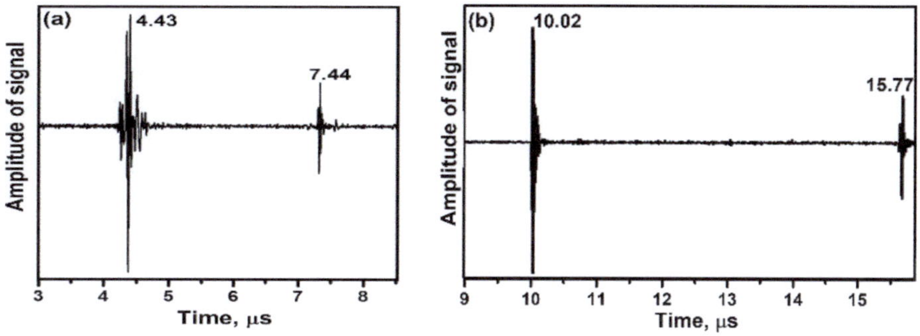

Fig. 4.7 Recorded signals extracted via **a** longitudinal waves and **b** transverse waves of $CaTiO_3$ specimen

a unit cell of $CaTiO_3$ is represented by the volume of a real crystal, so the unit cell is useful to acquire theoretical density. Nevertheless, calculations of planar density based on the unit cell were obtained, but with errors.

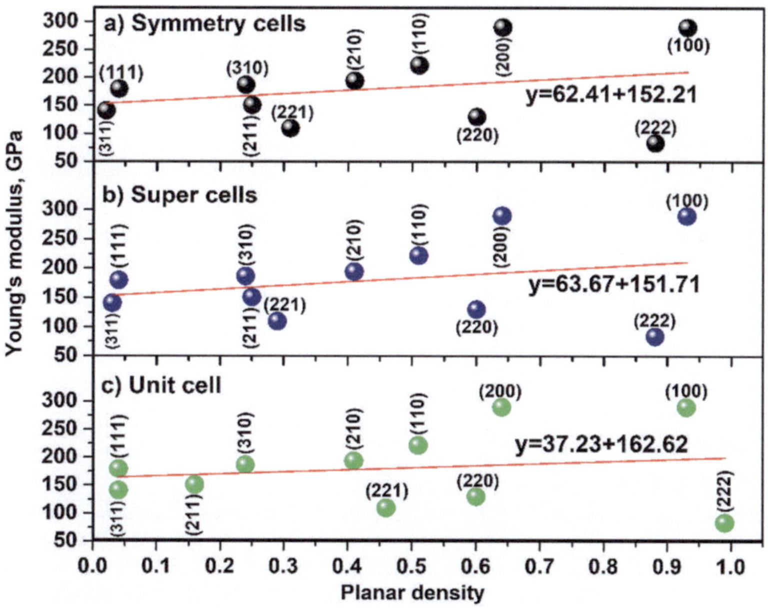

Fig. 4.8 Young's modulus versus planar density values of each diffracted plane related to the **a** symmetry cells, **b** super cells (2 × 2 × 2) and **c** unit cell of $CaTiO_3$

4.5 Conclusions

$CaTiO_3$ as a category of perovskite is successfully synthesized via solvothermal method.

Crystal size values of $CaTiO_3$ are calculated as ~59.10 and 63.02 through the Monshi-Scherrer method and BET analysis, and the crystal size values were confirmed by TEM image.

Planar density is responsible for modulus of elasticity of that plane; therefore, for the first time, comprehensive calculations of geometry, location and planar density values of $CaTiO_3$ were shown.

Elastic stiffness constants and Young's modulus values of $CaTiO_3$ were obtained by ultrasonic-echo method ($C_{11} = 330.89$, $C_{12} = 93.03$, $C_{44} = 94.91$ GPa and $E = 153.87 \pm 0.2$ GPa).

Young's modulus values of $CaTiO_3$ extracted by planar density and least square method were calculated as 162.62 ± 0.4, 151.71 ± 0.4 and 152.21 ± 0.4 GPa for unit cell, super cells ($2 \times 2 \times 2$) and symmetry cells, respectively.

The Young's modulus value of $CaTiO_3$ reported by symmetry cells is in good agreement with Young's modulus value reported by ultrasonic-echo technique and the literatre.

A unit cell of $CaTiO_3$ is not representative of the distribution of atoms on the planes; therefore, to obtain the real value of planar density and find the symmetry of distribution of atoms on the planes, expanded cells and utilizing symmetry cells are suggested.

Obtaining the planar density values based on unit cell or each super cells except for ($8 \times 8 \times 8$) is an estimation.

The real value of Young's modulus of $CaTiO_3$ should be extracted by symmetry cells or super cells ($8 \times 8 \times 8$).

The value of Young's modulus of $CaTiO_3$ extracted with this method can be applied for industrial applications.

References

1. M. Rabiei, A. Palevicius, S. Nasiri, A. Dashti, A. Vilkauskas, G. Janusas, Relationship between Young's modulus and planar density of unit cell, super cells (2 × 2 × 2), symmetry cells of Perovskite (CaTiO3) lattice. Materials. Basel: MDPI **14**(5), 1–15 art. no. 1258 (2021). ISSN 1996-1944. https://doi.org/10.3390/ma14051258. https://www.mdpi.com/1996-1944/14/5/1258
2. P.C. Reshmi Varma, Low-dimensional perovskites, in *Perovskite Photovoltaics: Basic to Advanced Concepts and Implementation*; Elsevier: Amsterdam, The Netherlands, pp. 197–229 (2018). ISBN 9780128129159
3. T. Nakamura, P.H. Sun, Y.J. Shan, Y. Inaguma, M. Itoh, I.N.S. Kim, J.H.O. Sohn, M. Ikeda, T. Kitamura, H. Konagaya, On the perovskite-related materials of high dielectric permittivity with small temperature dependence and low dielectric loss. Ferroelectrics **196**, 205–209 (1997)
4. X. Wang, C.N. Xu, H. Yamada, K. Nishikubo, X.G. Zheng, Electro-mechano-optical conversions in Pr^{3+}-doped $BaTiO_3$-$CaTiO_3$ ceramics. Adv. Mater. **17**, 1254–1258 (2005)

5. Y.Z. Ma, D. Sobernheim, J.R. Garzon, Glossary for unconventional oil and gas resource evaluation and development, in *Unconventional Oil and Gas Resources Handbook: Evaluation and Development*; Elsevier Inc.: Amsterdam, The Netherlands, pp. 513–526 (2016). ISBN 9780128025369
6. W.S. Slaughter, *The Linearized Theory of Elasticity*; Birkhäuser Boston: New York, NY, USA (2002)
7. S. Kuma, M.M. Woldemariam, Structural, electronic, lattice dynamic, and elastic properties of $SnTiO_3$ and $PbTiO_3$ using density functional theory. Adv. Condens. Matter Phys. **2019** (2019). https://doi.org/10.1155/2019/3176148
8. M. Nakamura, Elastic constants of some transition- metal- disilicide single crystals. Metall. Mater. Trans. A **25**, 331–340 (1994)
9. T. Paszkiewicz, S. Wolski, Elastic properties of cubic crystals: every's versus Blackman's diagram. J. Phys. Conf. Ser. **104**, 012038 (2008)
10. W.Y. Ching, P. Rulis, A. Misra, Ab initio elastic properties and tensile strength of crystalline hydroxyapatite. Acta Biomater. **5**, 3067–3075 (2009)
11. D. Holec, M. Friák, J. Neugebauer, P.H. Mayrhofer, Trends in the elastic response of binary early transition metal nitrides. Phys. Rev. B-Condens. Matter Mater. Phys. **85**, 64101 (2012)
12. K.M. Knowles, P.R. Howie, The directional dependence of elastic stiffness and compliance shear coefficients and shear moduli in cubic materials. J. Elast. **120**, 87–108 (2015)
13. K. Stair, J. Liu, M. Asta, Undefined Ultra-sonic measurement and computation of elastic constants, in *Proccedings of ASEE Annual Conference and Exposition*, Kansas City, MO, USA, 13–15 September (2006)
14. A. Kelly, K.M. Knowles, *Crystallography and Crystal Defects.* Wiley Chichester, UK (2012). ISBN 9781119961468
15. J. Epp, X-ray diffraction (XRD) techniques for materials characterization, in *Materials Characterization Using Nondestructive Evaluation (NDE) Methods.* Elsevier Inc.: Amsterdam, The Netherlands, pp. 81–124 (2016). ISBN 9780081000571
16. F. Badawi, P. Villain, Stress and elastic-constant analysis by X-ray diffraction in thin films. J. Appl. Crystallogr. **36**, 869–879 (2003)
17. Y. Waseda, E. Matsubara, K. Shinoda, *X-Ray Diffraction Crystallography X-ray Diffraction Crystallography Introduction, Examples and Solved Problems.* Springer: Berlin/Heidelberg, Germany (2011)
18. W. Dong, B. Song, W. Meng, G. Zhao, G. Han, A simple solvothermal process to synthesize $CaTiO_3$ microspheres and its-photocatalytic properties. Appl. Surf. Sci. **349**, 272–278 (2015)
19. Y. Wang, C.G. Niu, L. Wang, Y. Wang, X.G. Zhang, G.M. Zeng, Synthesis of fern-like Ag/AgCl/CaTiO3 plasmonic photocatalysts and their enhanced visible-light photocatalytic properties. RSC Adv. **6**, 47873–47882 (2016)
20. ASTM E797—95 standard practice for measuring thickness by manual ultrasonic pulse-echo contact method; ASTM international: West Conshohocken, PA, USA (2001)
21. ASTM E214—05 standard practice for immersed ultrasonic testing by the reflection method using pulsed longitudinal waves; ASTM international: West Conshohocken, PA, USA (2007)
22. S. Sahoo, S.K.S. Parashar, S.M. Ali, CaTiO3 nano ceramic for NTCR thermistor based sensor application. J. Adv. Ceram. **3**, 117–124 (2014)
23. E. Cockayne, B.P. Burton, Phonons and static dielectric constant in $CaTiO_3$ from first principles. Phys. Rev. B-Condens. Matter Mater. Phys. **62**, 3735–3743 (2000)
24. T. Janusas, S. Urbaite, A. Palevicius, S. Nasiri, G. Janusas, Biologically compatible lead-free piezoelectric composite for acoustophoresis based particle manipulation techniques. Sensors **21**, 483 (2021)
25. M. Rabiei, A. Palevicius, A. Monshi, S. Nasiri, A. Vilkauskas, G. Janusas, Comparing methods for calculating nano crystal size of natural hydroxyapatite using x-ray diffraction. Nanomaterials **10**, 1627 (2020)
26. Y. Li, R.B. Thompson, Relations between elastic constants Cij and texture parameters for hexagonal materials. J. Appl. Phys. **67**, 2663–2665 (1990)

27. H.B. Huntington, The elastic constants of crystals. Solid State Phys.-Adv. Res. Appl. **7**, 213–351 (1958)
28. M. De Jong, W. Chen, T. Angsten, A. Jain, R. Notestine, A. Gamst, M. Sluiter, C.K. Ande, S. Van Der Zwaag, J.J. Plata et al., Charting the complete elastic properties of inorganic crystalline compounds. Sci. Data **2**, 150009 (2015)
29. S. Tariq, A. Ahmed, S. Saad, S. Tariq, Structural, electronic and elastic properties of the cubic $CaTiO_3$ under pressure: a DFT study. AIP Adv. **5**, 077111 (2015)
30. A.P. Sakhya, J. Maibam, S. Saha, S. Chanda, A. Dutta, B.I. Sharma, R. Thapa, T. Sinha, Electronic structure and elastic properties of ATiO3 (A = Ba, Sr, Ca) perovskites: a first principles study. IJPAP **53**, 102–109 (2015)
31. H. Wang, P.L. Prendiville, P.J. McDonnell, W.V. Chang, An ultrasonic technique for the measurement of the elastic moduli of human cornea. J. Biomech. **29**, 1633–1636 (1996)
32. D.E. Bray, R.K. Stanley, *Nondestructive Evaluation: A Tool in Design, Manufacturing and Service; CRC Press: Boca Raton* (FL, USA, 1996)
33. R.S. Figliola, D.E. Beasley, *Theory and Design for Mechanical Measurements*, 7th edn. (John Wiley & Sons, Hoboken, NJ, USA, 2019)
34. M. Rabiei, A. Palevicius, A. Dashti, S. Nasiri, A. Monshi, A. Vilkauskas, G. Janusas, Measurement modulus of elasticity related to the atomic density of planes in unit cell of crystal lattices. Materials **13**, 4380 (2020)
35. M.A. Ramírez, R. Parra, M.M. Reboredo, J.A. Varela, M.S. Castro, L. Ramajo, Elastic modulus and hardness of $CaTiO_3$, $CaCu_3Ti_4O_{12}$ and $CaTiO_3/CaCu_3Ti_4O_{12}$ mixture. Mater. Lett. **64**, 1226–1228 (2010)
36. D.A. Papaconstantopoulos, M.J. Mehl, Tight-binding method in electronic structure, in *Encyclopedia of Condensed Matter Physics*; Elsevier Inc.: Amsterdam, The Netherlands, pp. 194–206 (2005). ISBN 9780123694010
37. J.M. Zhang, Y. Zhang, K.W. Xu, V. Ji, Young's modulus surface and Poisson's ratio curve for cubic metals. J. Phys. Chem. Solids **68**, 503–510 (2007)
38. W.P. Mason, Effect of dislocations on ultrasonic wave attenuation in metals. Bell Syst. Tech. J. **34**, 903–942 (1955)
39. R.E. Reed-Hill, R. Abbaschian, *Physical Metallurgy Principles*; Boston PWS publishing company: Boston, MA, USA (2009) ISBN 978-0-495-08254-5
40. V. Berdichevsky, Energy of dislocation networks. Int. J. Eng. Sci. **103**, 35–44 (2016)
41. D. Hull, D.J. Bacon, *Introduction to Dislocations*, 5th ed.; Butterworth-Heinemann: Burlington, MA, USA (2011). ISBN: 9780080966724
42. U. Englert, Symmetry relationships between crystal structures. Applications of crystallographic group theory in crystal chemistry. By Ulrich Müller. Angew. Chem. Int. Ed. **52**, 11973–11973 (2013)
43. O.H. Duparc, É. Polytechnique, O.B.M.H. Duparc, A review of some elements in the history of grain boundaries, centered on Georges Friedel, the coincident "site" lattice and the twin index crystallography history view project ANR FluTi view project a review of some elements in the history of grain boundaries, centered on Georges Friedel, the coincident "site" lattice and the twin index. Artic. J. Mater. Sci. **46**, 4116–4134 (2011)

Chapter 5
X-Ray Diffraction Analysis and the Williamson-Hall Method in the USDM Model to Estimate More Precise Values of Stress–Strain in the Unit Cell and Supercells (2 × 2 × 2) of Hydroxyapatite, as Validated by Ultrasonic Pulse-Echo Testing

This chapter illustrates the introduction of a novel technique to gauge the elastic modulus of crystallite structures like hexagonal HA [1]. One widely recognized approach to evaluate the stress–strain relationship in crystals is the Williamson-Hall (W–H) method, which incorporates X-Ray diffraction. The W–H method encompasses three models: (1) Uniform Deformation Model (UDM); (2) Uniform Stress Deformation Model (USDM); and (3) Uniform Deformation Energy Density Model (UDEDM). The models USDM and UDEDM have a direct connection with the modulus of elasticity (E). Young's modulus is crucial in engineering design and the development of materials. While Young's modulus is taken into account in the USDM and UDEDM models, historical studies have either used average values, calculated it from a singular XRD peak, or sourced it from literature, resulting in values not representative of all X-Ray diffraction peaks and lacking precision. However, the present study employs the W–H method by considering all diffracted planes of both the unit cell and 2 × 2 × 2 super cells of Hydroxyapatite (HA). By utilizing the USDM model with the W–H method, a highly accurate approach to determining stress (σ) and strain (ε) is introduced. A distinctive aspect of this research is its attention to the planar density of atoms. For verifying these novel assumptions, an ultrasonic pulse-echo test is also conducted.

Supplementary Information The online version contains supplementary material available at https://doi.org/10.1007/978-3-032-12583-5_5.

M. Rabiei et al., *Hydroxyapatite-Based Nanocomposites*, Springer Series in Chemical Physics 127, https://doi.org/10.1007/978-3-032-12583-5_5

5.1 Introduction

Young's modulus (E) is crucial for assessing mechanical characteristics, with precise knowledge of elastic modulus values being essential for comprehending material stiffness. Ceramic materials are increasingly used in industry due to their impressive mechanical properties, including high elastic moduli and hardness [2]. Hydroxyapatite, a notable bioceramic, has bioactive characteristics similar to natural bone minerals, offering significant biological importance [3]. Its chemical formula, Ca10(PO4)6(OH)2, closely resembles that of bone tissue [4, 5]. It is important to study the mechanical properties of hydroxyapatite during crystallization and development stages as Young's modulus directly influences crystal growth in mechanically stressed conditions [6]. Thus, focusing on the mechanical properties and structural design of hydroxyapatite is beneficial for both research and industrial purposes. Attention to its structural details is crucial for utilizing a simple, cost-efficient, and dependable method to measure Young's modulus. This study presents a novel approach that employs linear regression of the Young's modulus for each specific plane within a crystal lattice against planar density, aiming to accurately determine the overall Young's modulus for materials. In a pioneering effort, the planar density of diffracted planes of hydroxyapatite in both unit cells and super cells ($2 \times 2 \times 2$) was precisely calculated. Subsequently, we evaluated the comprehensive Young's modulus for the samples. Furthermore, we explored how cell size influences the Young's modulus by conducting an in-depth calculation of the planar density in hydroxyapatite super cells ($2 \times 2 \times 2$) and comparing it with data gathered from unit cell calculations. The Williamson-Hall (W–H) technique provides a method for analyzing stress and strain discovered through X-Ray diffraction. Moreover, the ultrasonic pulse-echo method is typically utilized for thickness measurement and defect sizing, reflecting signals from discontinuities within the test material's structure. *This research conducts a test to verify the validity of novelty hypotheses. The ultrasonic pulse echo setup, referred to as acoustic sound energy, identifies and localizes any defects or discontinuities. Moreover, ultrasonic pulse echo measurements are able to assess the elastic constant and compliance of materials* [7]. *The W–H method provides a fundamental calculation for plotting using XRD data. A significant issue in applying the W–H method within the USDM model is due to the Young's modulus values used in the equation. Often, these values are either taken from a single prominent peak in an XRD pattern or averaged values of Young's modulus, both of which are flawed as they do not accurately represent all diffracted planes. For instance, Ratan* et al. *calculated the stress (σ) and strain (ε) of cadmium selenide (CdSe) nanoparticles using the average Young's modulus from the Williamson-Hall (W–H) method within the USDM model* [8]. *Additionally, Khorsand* et al. *determined the Young's modulus of ZnO nanoparticles based on a prominent peak in the ZnO XRD pattern* [9]. *In another investigation, Rabiei* et al. *demonstrated the USDM model for Young's modulus in hydroxyapatite, taking into account the average across different diffracted planes [5Madhavi* et al. *applied the USDM model to determine the Young's modulus for VO2 doped ZnS/CdS composite nanopowder, utilizing average* *data for their calculations. Similarly, Rameshbabu* et al. *employed the W–H method combined with the USDM model to calculate the stress and strain values for hydroxyapatite, also using an average Young's modulus value* [10]. *Research involving the W–H method paired with the USDM model typically reports Young's modulus as an average or sourced from existing literature, lacking in high precision. Notably, values for the elastic compliance constant were also referenced from literature, while each material inherently possesses a unique elastic compliance constant; hence, relying on literature for these values without precise customization can impact accuracy. The W–H method is effective in assessing stress and strain of materials* [10]. *By employing our approach alongside the W–H method results, stress–strain estimates of samples can be achieved with enhanced precision. Our study utilized both DFT calculations and ultrasonic measurements to validate these findings (see Supplementary materials for details). The findings and derived values from this study are consistent with theoretical, experimental, and published data.*

5.2 Methods

The procedure for synthesizing hydroxyapatite powder is thoroughly detailed in Sect. 5.2 of the Supplementary Information (hydroxyapatite powder preparation).

5.2.1 Structural Analyses of Synthesized Hydroxyapatite

Figure 5.1 displays the XRD pattern of the synthesized hydroxyapatite powder. Several diffraction peaks in this pattern are consistent with hexagonal hydroxyapatite. The analysis was performed using X'pert, and the results aligned with the standard XRD peaks for hydroxyapatite (ICDD 9–432). Similar observations were reported in References [11, 12]. In addition, crystallographic parameters and details of each diffracted plane of hydroxyapatite were evaluated by X'pert and the values are tabulated in Tables 5.1 and S1.

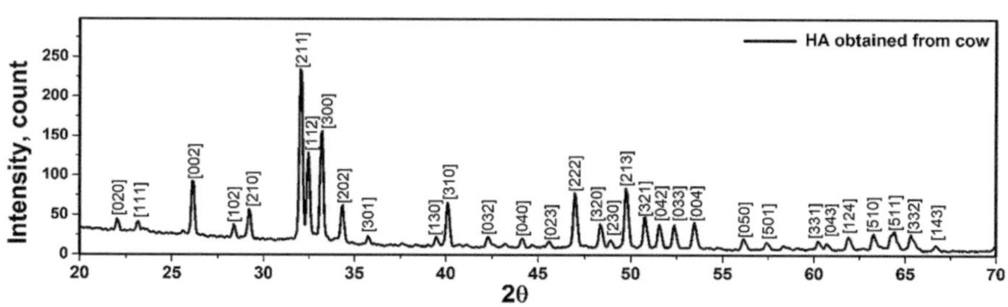

Fig. 5.1 X-Ray diffraction pattern of hydroxyapatite synthesized at 950 °C

Table 5.1 Crystallographic parameters of hydroxyapatite structure

Crystal system	a (Å)	c (Å)	Cell volume (Å) ^{3}	Crystal density (g/cm^3)	Space group
HCP	9.400	6.930	530.301	3.140	$P6_3/m$

According to Table S8.3.1, the values of the distance between planes are calculated by Eq. (1). In this equation, h, k and l are indices of each plane, and a, c and d are lattice parameters and distance of planes, respectively [13].

$$\frac{1}{d^2} = \frac{4}{3}\left(\frac{h^2 + hk + k^2}{a^2}\right) + \frac{l^2}{c^2} \tag{5.1}$$

Hydroxyapatite crystallizes in a hexagonal system, specifically within the P63/m space group, exhibiting minimal deviation from stoichiometry [14]. Figure 5.2 illustrates a unit cell diagram and includes a cif file of the synthesized hydroxyapatite. Within this structure, two types of calcium ion sites exist, leading to a total of 18 ions tightly forming the hexagonal arrangement. Each corner of the hexagon hosts a calcium ion, closely confined by 12 adjacent calcium ions, shared across three hexagons. Void regions between the hexagons accommodate three phosphate tetrahedra for each unit cell. This crystal matrix allows for the flexible substitution of its ions with biologically functional alternatives, a process known as doping, permitting the replacement of calcium, phosphate, and/or hydroxyl ions [15]. A distinctive characteristic of hydroxyapatite is the formation of, OH^- . ions into inner channels running parallel to the c axis, significantly influencing its mechanical and physical attributes [16]. Moreover, the Edax analysis for the synthesized hydroxyapatite is displayed in Fig. S8.3.2. The EDX analysis determined that the Ca/P ratio for the hydroxyapatite obtained from bovine bone is 1.60. Notably, throughout the sintering process, hydroxyapatite did not decompose thermally into tricalcium phosphate or tetra calcium phosphate, confirming successful synthesis, as supported by References [17, 18].

5.2.2 *Planar Density of Unit Cell and Super Cells (2 × 2 × 2) of Hydroxyapatite*

Based on the list of planes derived from X-Ray diffraction (see Fig. 5.1 and Table S8.3.1), Figs. 5.3, S8.3.6, and S8.3.7 depict the calculated planar density values for each diffracted plane within both the unit cell and super cells (2 × 2 × 2) of hydroxyapatite. The calculation of planar density values entails selecting diffracted planes sequentially from a low to a high angle. Importantly, all super cell lattices were assumed to be structured in a 2 × 2 × 2 matrix. The planar density is computed

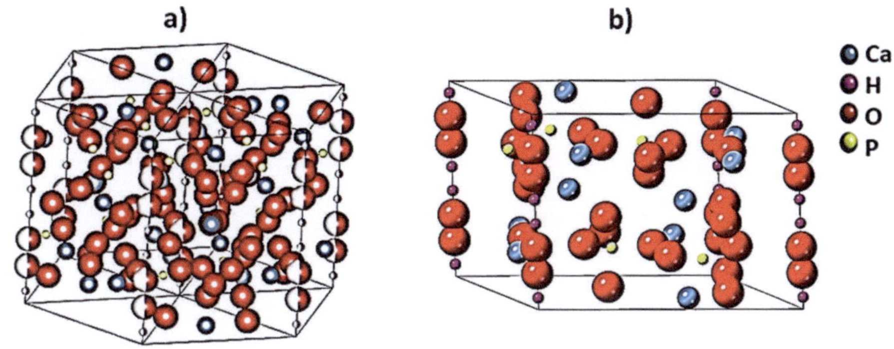

Fig. 5.2 Schematic representation of **a** hydroxyapatite unit cell and **b** the hydroxyapatite structure extracted by cif file

by dividing the area covered by atoms within the plane by the total area of the plane, using the atom centers as a reference [19].

5.2.3 *Elastic Stiffness Constant and Elastic Compliance of Hydroxyapatite*

Hooke's law is shown in Eq. (5.2); the stress corresponds to the strain for small displacements. It is the basic form, that this symmetry can be converted to the six items of stress (σ) and strain (ε) [20].

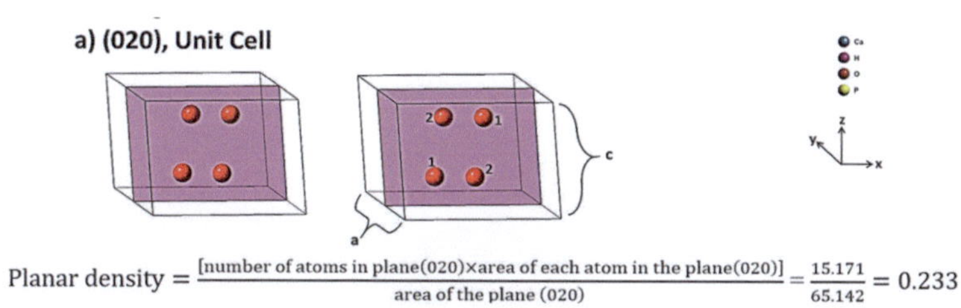

Fig. 5.3 Array and position of the involved atoms such as **a** (020) unit cell, **b** (020) super cell, **c** (111) unit cell, **d** (111) super cell, **e** (002) unit cell, **f** (002) super cell, **g** (102) unit cell and **h** (102) super cell (the calculations are in the Supplementary materials)

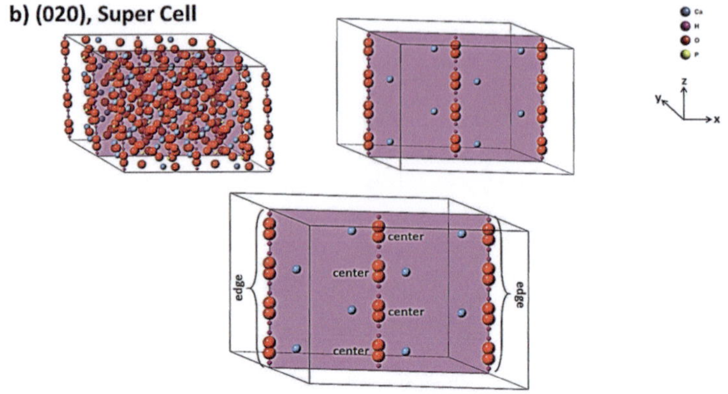

$$\text{Planar density} = \frac{[\text{number of atoms in plane}(020) \times \text{area of each atom in the plane }(020)]}{\text{area of the plane }(020)} = \frac{119.1299}{260.568} = 0.457$$

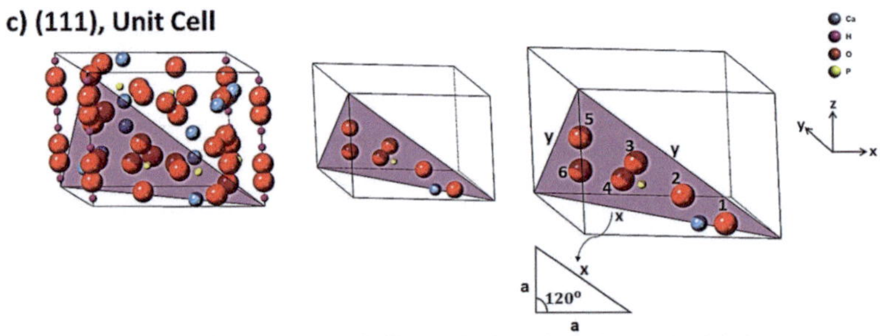

$$\text{Planar density} = \frac{[\text{number of atoms in plane}(111) \times \text{area of each atom in the plane}(111)]}{\text{area of the plane }(111)} = \frac{22.743}{68.171} = 0.334$$

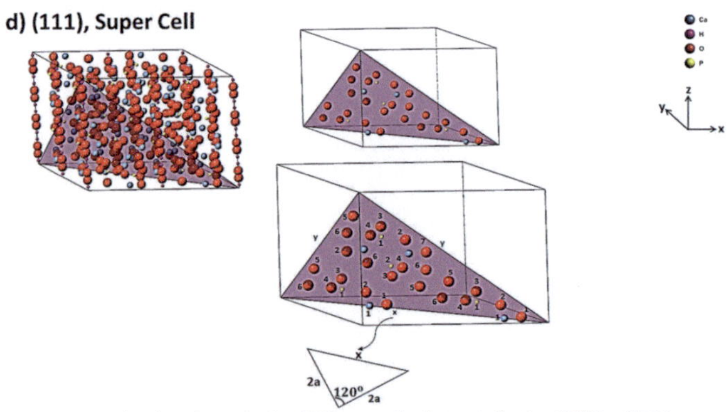

$$\text{Planar density} = \frac{[\text{number of atoms in plane}(111) \times \text{area of each atom in the plane}(111)]}{\text{area of the plane }(111)} = \frac{95.611}{272.704} = 0.351$$

Fig. 5.3 (continued)

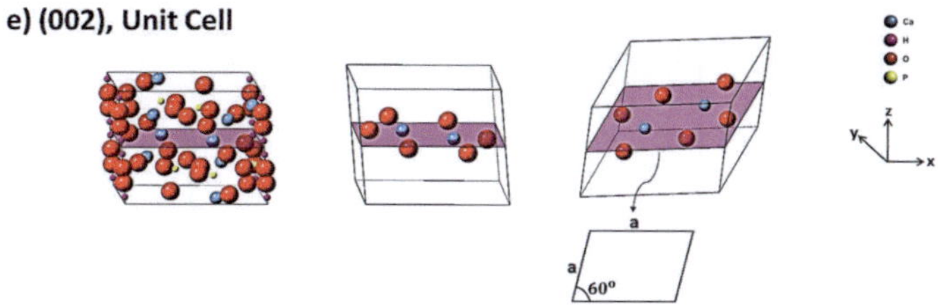

$$\text{Planar density} = \frac{[\text{number of atoms in plane(002)} \times \text{area of each atom in the plane(002)}]}{\text{area of the plane (002)}} = \frac{10.844}{76.522} = 0.142$$

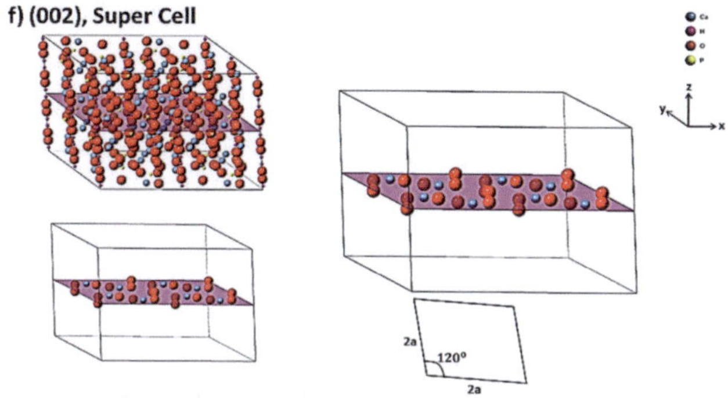

$$\text{Planar density} = \frac{[\text{number of atoms in plane(002)} \times \text{area of each atom in the plane(002)}]}{\text{area of the plane (002)}} = \frac{43.377}{306.88} = 0.142$$

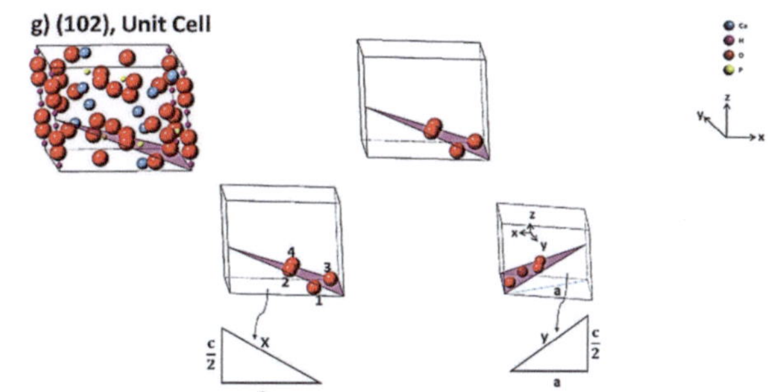

$$\text{Planar density} = \frac{[\text{number of atoms in plane(102)} \times \text{area of each atom in the plane(102)}]}{\text{area of the plane (102)}} = \frac{14.517}{41.581} = 0.349$$

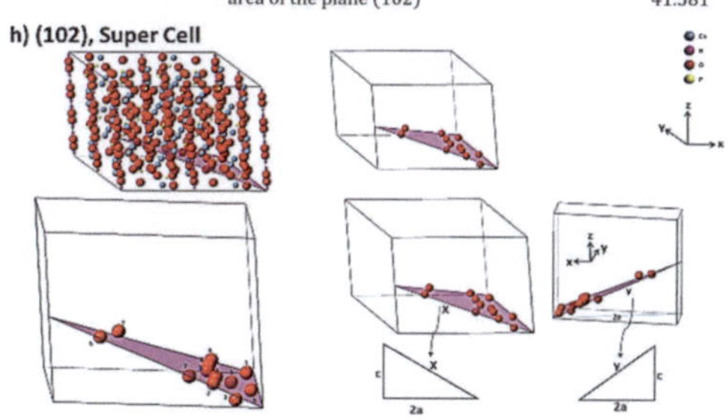

$$\text{Planar density} = \frac{[\text{number of atoms in plane(102)} \times \text{area of each atom in the plane(102)}]}{\text{area of the plane (102)}} = \frac{40.285}{166.367} = 0.242$$

Fig. 5.3 (continued)

$$\begin{pmatrix} \sigma_{xx} \\ \sigma_{yy} \\ \sigma_{zz} \\ \sigma_{yz} \\ \sigma_{zx} \\ \sigma_{xy} \end{pmatrix} = \begin{pmatrix} c_{11} & c_{12} & c_{13} & c_{14} & c_{15} & c_{16} \\ c_{21} & c_{22} & c_{23} & c_{24} & c_{25} & c_{26} \\ c_{31} & c_{32} & c_{33} & c_{34} & c_{35} & c_{36} \\ c_{41} & c_{42} & c_{43} & c_{44} & c_{45} & c_{46} \\ c_{51} & c_{52} & c_{53} & c_{54} & c_{55} & c_{56} \\ c_{61} & c_{62} & c_{63} & c_{64} & c_{65} & c_{66} \end{pmatrix} = \begin{pmatrix} \varepsilon_{xx} \\ \varepsilon_{yy} \\ \varepsilon_{zz} \\ \varepsilon_{yz} \\ \varepsilon_{zx} \\ \varepsilon_{xy} \end{pmatrix} \quad (5.2)$$

Additionally, Hooke's law can be written (Eq. (5.33)):

$$\begin{aligned} \sigma_{xx} &= C_{11}\varepsilon_{xx} + C_{12}\varepsilon_{yy} + C_{13}\varepsilon_{zz} + C_{14}\varepsilon_{yz} + C_{15}\varepsilon_{zx} + C_{16}\varepsilon_{xy} \\ \sigma_{yy} &= C_{21}\varepsilon_{xx} + C_{22}\varepsilon_{yy} + C_{23}\varepsilon_{zz} + C_{24}\varepsilon_{yz} + C_{25}\varepsilon_{zx} + C_{26}\varepsilon_{xy} \\ \sigma_{zz} &= C_{31}\varepsilon_{xx} + C_{32}\varepsilon_{yy} + C_{33}\varepsilon_{zz} + C_{34}\varepsilon_{yz} + C_{35}\varepsilon_{zx} + C_{36}\varepsilon_{xy} \\ \sigma_{yz} &= C_{41}\varepsilon_{xx} + C_{42}\varepsilon_{yy} + C_{43}\varepsilon_{zz} + C_{44}\varepsilon_{yz} + C_{45}\varepsilon_{zx} + C_{45}\varepsilon_{zx} \\ \sigma_{zx} &= C_{51}\varepsilon_{xx} + C_{52}\varepsilon_{yy} + C_{53}\varepsilon_{zz} + C_{54}\varepsilon_{yz} + C_{55}\varepsilon_{zx} + C_{56}\varepsilon_{xy} \\ \sigma_{xy} &= C_{61}\varepsilon_{xx} + C_{62}\varepsilon_{yy} + C_{63}\varepsilon_{zz} + C_{64}\varepsilon_{yz} + C_{65}\varepsilon_{zx} + C_{66}\varepsilon_{xy} \end{aligned} \quad (5.3)$$

The elastic stiffness determines the response of crystal to an externally applied stress or strain and provides information about bonding characteristics and mechanical and structural stability [21]. The hydroxyapatite system has five elastic constants (Eq. (5.4)). Therefore, the values of five independent C_{ij}, can be named C_{11}, C_{12}, C_{13}, C_{33}, C_{44}.

$$\text{Hydroxyapatitematrix} \begin{vmatrix} C_{11} & C_{12} & C_{13} & 0 & 0 & 0 \\ C_{12} & C_{11} & C_{13} & 0 & 0 & 0 \\ C_{13} & C_{13} & C_{33} & 0 & 0 & 0 \\ 0 & 0 & 0 & C_{44} & 0 & 0 \\ 0 & 0 & 0 & 0 & C_{44} & 0 \\ 0 & 0 & 0 & 0 & 0 & \frac{1}{2}(C_{11} - C_{12}) \end{vmatrix} \quad (5.4)$$

The crystallographic nature of the hexagonal structure is shown in Figure S3. Furthermore, for conventional hexagonal systems, such as hydroxyapatite, the relationship between C_{ij} and S_{ij} is introduced in Eqs. (5.5–5.9) [22, 23].

$$S_{11} = \frac{1}{2}\left(\frac{C_{33}}{C_{33}(C_{11} + C_{12}) - 2(C_{13})^2} + \frac{1}{C_{11} - C_{12}}\right) \quad (5.5)$$

$$S_{12} = \frac{1}{2}\left(\frac{C_{33}}{C_{33}(C_{11} + C_{12}) - 2(C_{13})^2} - \frac{1}{C_{11} - C_{12}}\right) \quad (5.6)$$

$$S_{33} = \frac{C_{11} + C_{12}}{C_{33}(C_{11} + C_{12}) - 2(C_{13})^2} \quad (5.7)$$

$$S_{13} = -\frac{C_{13}}{C_{33}(C_{11} + C_{12}) - 2(C_{13})^2} \quad (5.8)$$

$$S_{44} = \frac{1}{C_{44}} \tag{5.9}$$

Based on Eqs. (5.5–5.9), it is necessary to determine C values in order to derive S values. We employed two methods to ascertain the C values. The first method involved performing theoretical calculations using the CASTEP model within the materials studio software version 6.0, utilizing the GGA level of theory with a PBE basis set (as illustrated in Fig. S8.3.4). The second method relied on ultrasonic measurements (presented in Table S8.3.2). Through these ultrasonic measurements of phase velocity anisotropy, we determined the complete set of five elastic stiffness constants (C11, C12, C13, C33, and C44) for the samples. The stiffness constants were documented by applying Eqs. (5.10–5.15), wherein ρ represents the sample's density and V indicates velocity [24–28].

$$C_{11} = \rho V^2_{1/1}, C_{22} = \rho V^2_{2/2} \tag{5.10}$$

$$C_{66} = \rho V^2_{1/2} = \rho V^2_{2/1}, C_{55} = \rho V^2_{1/3} = \rho V^2_{3/1} \tag{5.11}$$

$$C_{12} = \sqrt{\left(C_{11} + C_{66} - 2\rho V^2_{12/12}\right)\left(C_{22} + C_{66} - 2\rho V^2_{12/12}\right)} - C_{66} \tag{5.12}$$

$$C_{44} = \rho V^2_{2/3} = \rho V^2_{3/2} \tag{5.13}$$

$$C_{13} = \sqrt{\left(C_{11} + C_{55} - 2\rho V^2_{13/13}\right)\left(C_{33} + C_{55} - 2\rho V^2_{13/13}\right)} - C_{55} \tag{5.14}$$

$$C_{33} = \rho V^2_{3/3} \tag{5.15}$$

To measure velocities, the standard ultrasonic pulse-echo method ASTM E797/E797-M-15 was employed, as referenced in [26]. The setup involved immersing the probe and sample in water to minimize pressure effects caused by a hand-held probe, and a 5.4 MHz longitudinal frequency probe was used. To reduce wave extension and energy loss, a lens with specific curvature was applied according to the guidelines of Reference [28]. To determine the C66 value, high amplitude curves were generated by altering rotation angles. For generating transverse waves, a 2.3 MHz probe and a viscous interface material (honey) were utilized, with sample pressure adjusted for optimal and smoother curve results, as detailed in [29]. Taking into account each point of the samples, a three-dimensional axis, such as X_1, X_2 and X_3, can be performed. In this case, according to Fig. S5, X_1 is the radial coordinate, X_2 is the circumferential coordinate and X_3 is the axial coordinate. $V_{i/j}$ is the denoted velocity of an ultrasound wave that can be propagated in the X_i direction with particle displacements in the Xj direction simultaneously. $V_{i/j}$,, with the same i and j, is longitudinal and with i ≠ j being transverse waves. For measuring quasi-longitudinal or quasi-transverse velocity (Vij/ij), specimens should be cut (bezel) on the edges toward surfaces of perpendicular X directions (Fig/. S5). The obtained values of velocities are shown in Table S8.3.2. On the other hand, C_{11} is in good agreement with the longitudinal distortion and longitudinal compression/tension; thus, C_{11} can be introduced as the hardness. Moreover, the transverse distortion depends on the C_{12}, and C_{12} comes from the transverse expansion related to the Poisson's ratio. Additionally, C_{44} corresponds to the shear modulus, and C_{44} is in the settlement with C_{11} and C_{12} [22, 30]. Accordingly, the shear modulus is proportional to the Burgers vector and the Young's modulus; in addition, dislocation density is in agreement with the Young's modulus [31, 32].

In the ultrasonic method, longitudinal and transverse waves were utilized for measuring the Young's modulus value [33, 34]. According to this method (Eq. (16)), based on the velocity of ultrasound waves and density of sample, the Young's modulus value was determined.

$$E = \frac{\rho c_l^2 \left[3\left(\frac{c_l}{c_t}\right)^2 - 4\right]}{\left(\frac{c_l}{c_t}\right)^2 - 1} \tag{5.16}$$

In Eq. (5.16), ρ, c_l and c_t are density, velocity of longitudinal and transverse ultrasound waves tandemly. Furthermore, according to Eq. (5.17), the velocity of longitudinal and transverse waves can be registered by determining the length of specimen and the differences between two echoes ($t = t_2 - t_1$) in the signals [35].

$$c = \frac{2L}{t} \tag{5.17}$$

Here, L is the length of the sample and t is the difference between two echoes, and the density of the sample can be detected by measuring the mass and volume of the sample [36]. Additionally, with the substitution of Eqs. (5.16) and (5.17), the main equation for the calculation of Young's modulus is Eq. (5.18).

$$E = \frac{4\rho\left(\frac{L}{t_s}\right)^2 \left(3t_s^2 - 4t_l^2\right)}{t_s^2 - t_l^2} \tag{5.18}$$

t_s and t_l are differences between two echoes in longitudinal and transverse waves separately [37]. Table 5.2 provides a comparison of the theoretical and experimental results for the elastic stiffness constants of hydroxyapatite, utilizing both previously published data and findings from this research. Furthermore, Table 5.3 lists the elastic compliance values that were determined here, based on Eqs. (5.10–5.15). As a result, the theoretical values show good agreement with those from References [38, 39]. Table 5.4 also includes the repeated measurements (conducted five times) for the times of transverse and longitudinal wave propagation within the samples.

Table. 5.2 Theoretical and experimental values of the Elastic constant of hydroxyapatite

Stiffness constant (C), (Gpa)	Theory from ref [40] (Ching et al.)	Theory from ref [41] (Leeuw et al.)	Experiment. from ref [42] (Katz et al.)	Theory this study	Experiment this study
C_{11}	140.00	134.40	137.00	139.58	135.78
C_{12}	42.40	48.90	42.50	48.03	49.21
C_{13}	58.30	68.50	54.90	61.22	56.62
C_{33}	174.80	184.70	172.00	181.08	179.22
C_{44}	47.50	51.40	39.60	50.93	41.73

Table. 5.3 Theoretical and experimental values of the Elastic compliance of hydroxyapatite

Compliance constant (S), (Gpa)	Theory (Ching et al.)	Theory (Leeuw et al.)	Experiment (Katz et al.)	Theory this study	Experiment this study
S_{11}	0.008607002	0.009621806	0.008752330	0.008881114	0.009126549
S_{12}	–0.001638900	–0.002074100	–0.001829681	–0.002041879	–0.002424797
S_{13}	–0.002324030	–0.002799231	–0.002209613	–0.002312227	–0.002117248
S_{33}	0.007271063	0.007490496	0.007224509	0.007085868	0.006917516
S_{44}	0.021052632	0.019455253	0.025252525	0.019634793	0.023963575

Table 5.4 Resulted values of time for transverse and longitudinal waves of samples

Number of measurement	t_s (μs)	t_l (μs)	L (mm)
1	6.26	2.52	11.59
2	6.27	2.50	11.61
3	6.26	2.51	11.59
4	6.25	2.53	11.57
5	6.24	2.54	11.55

5.2.4 *The Young's Modulus Versus Planar Density of Unit Cell and Super Cells (2 × 2 × 2) of Hydroxyapatite*

X-Ray diffraction has provided data on diffracted planes and the location of atoms in each plane. In the previous section, the planar density of each diffracted plane was calculated and it could play an important role in the mechanical properties of each plane. The Young's modulus of each plane (E_{hkl}) of a hydroxyapatite lattice can be calculated as Eq. (5.19) [5, 43]. In this equation, h, k and l are the plane indices, a and c are the lattice parameters, C and S are the elastic stiffness constant and elastic compliance, respectively. The values of Young's modulus of 32 diffracted planes of hydroxyapatite in unit cells and super cells (2 × 2 × 2),

Table 5.5 Young's modulus values of unit cell and super cell lattices of hydroxyapatite

Study	Young Modulus (E), (Gpa) in This method (Intercept Value)	
	Unit cell	Super cells (2 × 2 × 2)
Theory. (Ching et al.)	118.43	127.51
Theory. (Leeuw et al.)	113.94	128.91
Experiment. (Katz et al.)	109.54	117.52
Theory. (This study)	118.62	131.74
Experiment. (This study)	108.15	121.17

$E_{(h_1k_1l_1)}, E_{(h_2k_2l_2)}, E_{(h_3k_3l_3)} \ldots\ldots\ldots\ldots, E_{(h_{32}k_{32}l_{32})}$, related to the literature and the present study, are reported in Table S8.3.3.

$$E_{hkl} = \frac{\left[h^2 + \frac{(h+2k)^2}{3} + \left(\frac{al}{c}\right)^2\right]^2}{S_{11}\left(h^2 + \frac{(h+2k)^2}{3}\right)^2 + S_{33}\left(\frac{al}{c}\right)^4 + (2S_{13} + S_{44})\left(h^2 + \frac{(h+2k)^2}{3}\right)\left(\frac{al}{c}\right)^2} \quad (5.19)$$

Utilizing the least squares technique to analyze the relationship between Young's modulus and the planar density of diffracted planes (following our proposed approach), we accurately determined the Young's modulus of hydroxyapatite. Therefore, calculating the C and S parameters for crystalline materials is crucial for implementing this method. To demonstrate the practicality and accuracy of our proposed technique for assessing Young's modulus, Fig. 4 presents the Young's modulus values for each plane relative to the planar density of both the unit cell and supercells (2 × 2 × 2). The intercept values for the Young's modulus of hydroxyapatite's unit cell and supercells (2 × 2 × 2), derived from the least squares method, are documented in Table 5.5.

Utilizing uncertainty measurements (conducted in five replications per Table 5.4) and refer to Eq. (5.18), the Young's modulus was found to be 113.08 ± 0.14 GPa through ultrasonic methods. This measurement aligns well with our study's reported values in Table 5.5. The discrepancy between theoretical predictions and experimental outcomes for both the unit cell and 2 × 2 × 2 super cells is consistent, standing at 10.47 GPa and 10.57 GPa, respectively. This suggests that the theoretical model is accurate, and by subtracting approximately 10 GPa, the actual experimental results can be achieved.

5.3 Result and Discussion

5.3.1 *Positive and Negative Slope Values of Unit Cell and Super Cells (2 x 2 x 2)*

The variation in Young's modulus between the unit cell and super cells (2 × 2 × 2) can be traced to the atomic positioning. For the super cells (2 × 2 × 2), a negative calculated slope is observed. Initially, it is evident that this slope is influenced by the planar density of the diffracted planes. Therefore, a fitting based on the planar density of super cells, composed of a matrix of eight unit cells, may yield a more accurate intercept result [35]. This occurs because the arrangement of eight cells together offers enhanced symmetry compared to the configuration of two cells (Fig. S8.3.8) [35]. Additionally, the difference in slope, being positive in the unit cell and negative in the super cells (2 × 2 × 2), can be attributed to defects like point imperfections (vacancies, substitutions, and interstitials), line imperfections (screw and edge dislocations), surface imperfections (grain boundaries), and volume defects (disorder in atomic arrangement within tiny amorphous regions). These imperfections exert a greater effect in the super cells (2 × 2 × 2), whereas the unit cell remains relatively ideal and less impacted by these imperfections. As the density of atoms across planes increases, it reduces the force needed for dislocation motion, thus decreasing strength, which ultimately results in a negative slope for the super cells (2 × 2 × 22). For super cells, increasing the number of atoms enhances planar density, which in turn heightens the effect of dislocation motion. This causes a reduction in both strength and Young's modulus, resulting in a negative slope. The intercept of the fitted line gives a value of Young's modulus, representing the modulus of a plane with zero planar density, as if the plane lacks any particular atom. Hence, discussions regarding defects and imperfections are not applicable to such a plane, as there is no atom at its origin. Consequently, the Young's modulus of the unit cell is lower compared to that of the super cells (2 × 2 × 2). As a result, in a smaller unit cell area, the plane without any atom is more realistically represented than in the larger area of the super cells. The intercept within a unit cell is thus more accurately aligned with the real Young's modulus values. As previously discussed, the planar density is influenced by both the arrangement and positioning of atoms within a plane, as well as the orientation of these planes. For example, as shown in Fig. 5.4b, the planar density relationship of (210) > (510) > (102) > (023) > (043) arises because the order of planes does not reflect diffraction (XRD) but rather relies on the planar density values. This approach enables the comparison of two or more planes with similar planar density values; for instance, in supercells (2 × 2 × 2), the planar density of both (040) and (331) equals 0.250. The displacement and deformation processes of atoms in the planes are linked to dislocation networks. The work or energy (W) required for atomic movement within each plane is expressed in Eq. (20) [44, 45].

$$W = Gb^2 1 \tag{5.20}$$

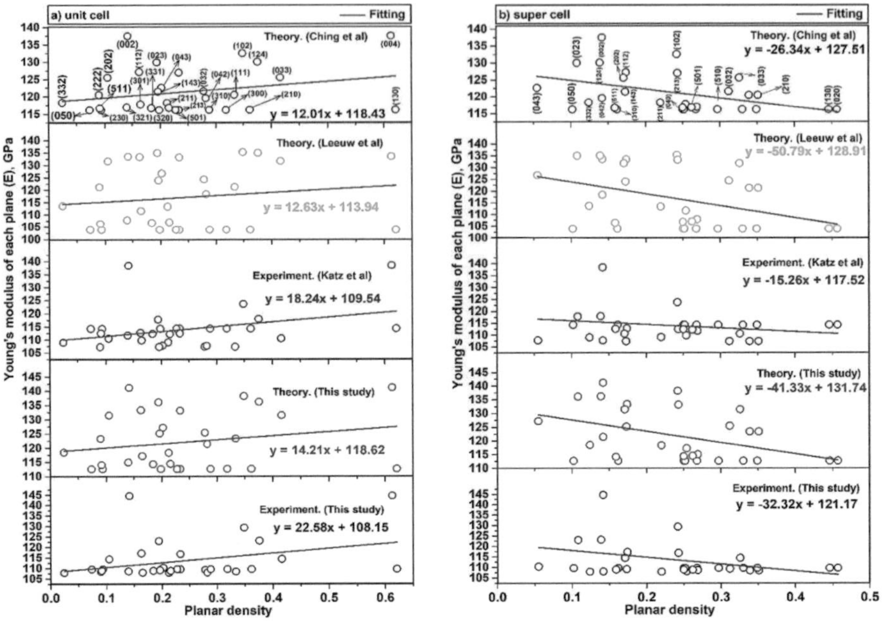

Fig. 5.4 Young's modulus of each plane **a** unit-cell **b** super cells (2 × 2 × 2) of hydroxyapatite extracted by XRD patterns and planar density

In this equation, G, b and l are shear modulus, Burgers vector and dislocation length, respectively. In addition, by merging Eqs. (5.20) and (5.21), Eq. (5.22) can be registered. In Eqs. (5.21) and (5.22), E and ν are Young's modulus and Poisson's ratio tandemly.

$$G = \frac{E}{2(1 + \nu)} \tag{5.21}$$

$$W = \frac{E}{2(1 + \nu)} b^2 1 \tag{5.22}$$

Multiple studies have examined the characteristics of hydroxyapatite, with a particular focus on its biocompatibility and bioactivity, which rely on the understanding and identification of hydroxyapatite structures [46, 47]. For instance, when atoms are distantly spaced like the corner atoms (Ca) in (111) super cells (2 × 2 × 2), they necessitate greater applied force to be brought closer together. Understanding the planes of hydroxyapatite structure is therefore advantageous for metal, polymer, and ceramic doping, facilitating processes like controlled protein release and crafting bioactive monolithic fragments in the biomaterials sector [48, 49]. Furthermore, the hydroxyl ion resides at the center of each unit cell. It is circled by three calcium ions per hexagon in the center of the hydroxyapatite lattice, forming six-ion rings. This structure, forming a chord, is pivotal to many hydroxyapatite properties, especially its biocompatibility; the positioning of the hydroxyl group within the planes plays a vital role in doping processes [15, 50].

5.3.2 Williamson-Hall Method in USDM Model

The W–H method has been used for determining different elastic properties. The best procedure is to mathematically reduce the errors and obtain the values of the Young's modulus by all the diffracted peaks, using the least squares method. The ***W–H method is a simplified integral expanse and, taking into account the peak width, strain-induced broadening is specified*** [51]***. Taking into account the W–H method in the USDM model,*** Young's modulus values are examined. It is clear that in Eq. (5.23) and Fig. 5, the term of $\frac{4\sin\theta}{E_{hkl}}$ is along the X-axis and the term of $\beta_{hkl}.\cos\theta$ is along the Y-axis individually.

$$\beta_{hkl}\cdot\cos\theta = \left(\frac{K\lambda}{L}\right) + 4\sigma\frac{\sin\theta}{E_{hkl}} \tag{5.23}$$

Here, β_{hkl} is the broadening peak from (hkl) plane and, in this study, the instrumental broadening is taken as 0.02 degree for each diffracted peak and is subtracted from β_{hkl}. values before multiplying to $\frac{\pi}{180}$ to convert the degree to radian. In this model, the condition is performed to calculate the strain and the average Young's modulus. The calculated average E value within studies employinthe W–H method is prone to errors. This is because when considering the average Young's modulus values through Eq. (5.19), the result significantly diverges from the standard Young's modulus for each X-ray diffraction-extracted peak. Furthermore, some studies reference Young's modulus values from literature that do not pertain to the materials being analyzed. For example, as highlighted in Reference [10], which examines the W–H method's application for determining crystal size and performing Scherrer analysis on hydroxyapatite, the average Young's modulus value shows a more significant deviation compared to the actual Young's modulus of hydroxyapatite. The line broadening of the diffracted peak is determined using Crystal Diffract 6.7.2.300 software. Figure 5.5 illustrates slope values relating to stress (σ), with positive slope values indicating tensile stress and intrinsic strain, while negative values would suggest compressive stress and strain. Moreover, the stress (σ) and strain (ε) outcomes from the W–H method in the USDM model are presented in Table 5.6, with σ values aligning closely with those in Reference [52].

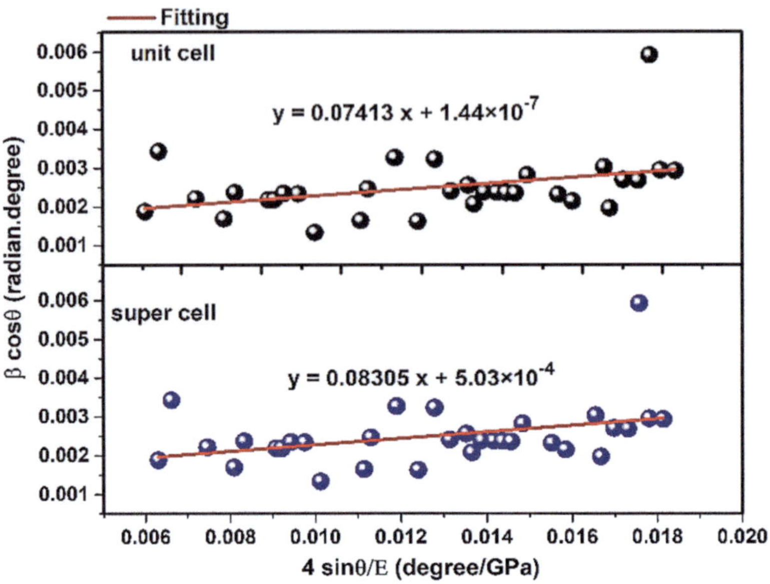

Fig. 5.5 W–H in USDM model and plot of unit cell and super cells (2 × 2 × 2) of hydroxyapatite

Table. 5.6 Mechanical properties values of unit cell and super cells (2 × 2 × 2) of hydroxyapatite

Mechanical Properties			
Structure	σ (GPa)	ε	E[1] (GPa)
Unit cell	0.07413	0.00068	108.15
Super cells (2 × 2 × 2)	0.08305	0.00068	121.17

[1]Experiments of Young's modulus values (This Study)

5.4 Conclusions

In this investigation, hydroxyapatite crystals were effectively synthesized using thermal treatment. Additionally, for the first time, the atomic positions and the calculated planar density values of 32 diffracted planes within the hydroxyapatite structure were determined. Furthermore, a novel approach was introduced that explores the connection between measuring the elastic modulus and atomic planar density in crystalline hydroxyapatite, encompassing both a unit cell and super cells (2 × 2 × 2). This study revealed that the slope of the elastic modulus versus planar density for super cells (2 × 2 × 2) is negative. This finding is linked to structural imperfections, suggesting that increased atomic density in planes results in a reduced requirement for dislocation motion forces. Consequently, a plane devoid of atoms is more realistic in smaller unit cell areas compared to the larger areas in super cells (2 × 2 × 2). This implies that the intercept related to the unit cell provides a more accurate representation of Young's modulus in the hydroxyapatite lattice. In addition, comparing the theoretical and experimental data for the Young's modulus of hydroxyapatite reveals a minor difference in values for both unit cells and supercells (2 × 2 × 2), with recorded values of 10.47 GPa and 10.57 GPa, respectively. This indicates that the theoretical prediction is accurate, and the experimental value can be achieved by a reduction of roughly 10 GPa. Also, the Young's modulus for hydroxyapatite for the unit cell and supercells (2 × 2 × 2) were determined to be 108.15 GPa and 121.17 GPa, respectively. Lastly, one application of the discussed method was implemented in this study, showing that the Williamson-Hall method within the USDM framework can be used to reduce errors in the least squares method, yielding a precise elastic modulus for hydroxyapatite that surpasses the average calculated value in accuracy.

References

1. M. Rabiei, A. Palevicius, S. Nasiri, A. Dashti, A Vilkauskas, G. Janusas, Relationship between Young's modulus and planar density of unit cell, super cells (2 × 2 × 2), symmetry cells of Perovskite (CaTiO3) lattice. Materials. Basel: MDPI 2021 **14**(5), 1–15, art. no. 1258 (202), ISSN 1996-1944. https://doi.org/10.3390/ma14051258 https://www.mdpi.com/1996-1944/14/5/1258
2. M. Shimada, K. Matsushita, S. Kuratani, T. Okamoto, M. Koizumi, K. Tsukuma, T. Tsukidate, Temperature dependence of young's modulus and internal friction in alumina, silicon nitride, and partially stabilized zirconia ceramics. J. Am. Ceram. Soc. **67**, C23–C24 (1984). https://doi.org/10.1111/j.1151-2916.1984.tb09612.x
3. M. Vallet-Regí, Revisiting ceramics for medical applications. Dalton Trans. 5211–5220 (2006). https://doi.org/10.1039/b610219k
4. R. Marzieh, N. Sohrab, P. Arvydas, J. Giedrius, Preparation and investigation of bioactive organic-inorganic nano-composite derived from PVB-co-VA-co-VAc/HA, in *Proceedings of the 15th International Conference Mechatronic Systems and Materials, MSM 2020* (Bialystok, Poland, 1–3 July 2020 Institute of Electrical and Electronics Engineers Inc.: Piscataway, NJ, USA, 2020)
5. M. Rabiei, A. Palevicius, A. Monshi, S. Nasiri, A. Vilkauskas, G. Janusas, Comparing methods for calculating nano crystal size of natural hydroxyapatite using X-ray diffraction. Nanomaterials **10**, 1627 (2020). https://doi.org/10.3390/nano10091627
6. W.J. Shih, M.C. Wang, M.H. Hon, Morphology and crystallinity of the nanosized hydroxyapatite synthesized by hydrolysis using cetyltrimethylammonium bromide (CTAB) as a surfactant. J. Cryst. Growth **275**, e2339–e2344 (2005)
7. T.M. Loganathan, M.T.H. Sultan, M.K. Gobalakrishnan, Ultrasonic inspection of natural fiber-reinforced composites. In *Sustainable Composites for Aerospace Applications* (Elsevier: Amsterdam, The Netherlands, 2018), pp. 227–251, ISBN 9780081021316
8. D. Nath, F. Singh, R. Das, X-Ray diffraction analysis by Williamson-Hall, Halder-Wagner and size-strain plot methods of CdSe nanoparticles—a comparative study. Mater. Chem. Phys. **239**, 122021 (2020). https://doi.org/10.1016/j.matchemphys.2019.122021
9. A. Khorsand Zak, W.H. Abd Majid, M.E. Abrishami, R. Yousefi, X-Ray analysis of ZnO nanoparticles by Williamson-Hall and size-strain plot methods. Solid State Sci. **13**, 251–256 (2011). https://doi.org/10.1016/j.solidstatesciences.2010.11.024
10. K. Venkateswarlu, A. Chandra Bose, N. Rameshbabu, X-ray peak broadening studies of nanocrystalline hydroxyapatite by WilliamsonHall analysis. Phys. B Condens. Matter **405**, 4256–4261 (2010). https://doi.org/10.1016/j.physb.2010.07.020

11. M. Bahrololoom, M. Javidi, S. Javadpour, Characterisation of natural hydroxyapatite extracted from bovine cortical bone ash. J. Ceram. Process. Res. **10**, 129–138 (2009)
12. S. Shahabi, F. Najafi, A. Majdabadi, T. Hooshmand, M. Haghbin Nazarpak, B. Karimi, S.M. Fatemi, Effect of gamma irradiation on structural and biological properties of a PLGA-PEG-hydroxyapatite composite. Sci. World J. **2014** (2014). https://doi.org/10.1155/2014/420616
13. M. Pasero, A.R. Kampf, C. Ferraris, I.V. Pekov, J. Rakovan, T.J. White, Nomenclature of the apatite supergroup minerals. Eur. J. Mineral. **22**, 163–179 (2010). https://doi.org/10.1127/0935-1221/2010/0022-2022
14. L.L. Hench, *An Introduction to Bioceramics*, 2nd ed. (Imperial College Press London, UK, 2013), ISBN 9781908977168
15. K. Lin, J. Chang, Structure and properties of hydroxyapatite for biomedical applications, in *Hydroxyapatite (Hap) for Biomedical Applications* (Elsevier: Amsterdam, The Netherlands, 2015), pp. 3–19
16. V.S. Bystrov, J. Coutinho, A.V. Bystrova, Y.D. Dekhtyar, R.C. Pullar, A. Poronin, E. Palcevskis, A. Dindune, B. Alkan, C. Durucan et al., Computational study of hydroxyapatite structures, properties and defects. J. Phys. D Appl. Phys. **48**, 195302 (2015). https://doi.org/10.1088/0022-3727/48/19/195302
17. M. Rajkumar, N. Sundaram, V. Rajendran, Preparation of size controlled, stoichiometric and bioresorbable hydroxyapatite nanorod by varying initial pH, Ca/P ratio and sintering temperature. Digest J. Nanomater. Biostruct. **6**, 169–179 (2011)
18. E. Landi, S. Riccobelli, N. Sangiorgi, A. Sanson, F. Doghieri, F. Miccio, Porous apatites as novel high temperature sorbents for carbon dioxide. Chem. Eng. J. **254**, 586–596 (2014). https://doi.org/10.1016/j.cej.2014.05.070
19. M. Rabiei, A. Palevicius, A. Dashti, S. Nasiri, A. Monshi, A. Vilkauskas, G. Janusas, Measurement Modulus of elasticity related to the atomic density of planes in unit cell of crystal lattices. Materials **13**, 4380 (2020). https://doi.org/10.3390/ma13194380
20. K. Rajabi, S. Hosseini-Hashemi, Application of the generalized Hooke's law for viscoelastic materials (GHVMs) in nanoscale mass sensing applications of viscoelastic nanoplates: a theoretical study. Eur. J. Mech. A. Solids **67**, 71–83 (2018). https://doi.org/10.1016/j.euromechsol.2017.08.015
21. M.B. Kanoun, S. Goumri-Said, K. Abdullah, Theoretical study of physical properties and oxygen incorporation effect in nanolaminated ternary carbides 211-MAX phases, in *Advances in Science and Technology of Mn+1AXn Phases* (Elsevier: Amsterdam, The Netherlands, 2012), pp. 177–196
22. Y. Li, R.B. Thompson, Relations between elastic constants Cij and texture parameters for hexagonal materials. J. Appl. Phys. **67**, 2663–2665 (1990). https://doi.org/10.1063/1.345479
23. H.B. Huntington, The elastic constants of crystals. Solid State Phys. Adv. Res. Appl. **7**, 213–351 (1958). https://doi.org/10.1016/S0081-1947(08)60553-6
24. M. Mah, D.R. Schmitt, Determination of the complete elastic stiffnesses from ultrasonic phase velocity measurements. J. Geophys. Res. Solid Earth **108** (2003). https://doi.org/10.1029/2001jb001586
25. J.R. Neighbours, G.E. Schacher, Determination of elastic constants from sound-velocity measurements in crystals of general symmetry. J. Appl. Phys. **38**, 5366–5375 (1967). https://doi.org/10.1063/1.1709328
26. ASTM E797/E797M-15, Standard Practice for Measuring Thickness by Manual Ultrasonic Pulse-Echo Contact Method. https://www.astm.org/Standards/E797.htm. Accessed 19 Nov 2020
27. W.C. Van Buskirk, S.C. Cowin, R.N. Ward, Ultrasonic measurement of orthotropic elastic constants of bovine femoral bone. J. Biomech. Eng. **103**, 67–72 (1981). https://doi.org/10.1115/1.3138262
28. ASTM E214-05, Standard Practice for Immersed Ultrasonic Testing by the Reflection Method Using Pulsed Longitudinal Waves (Withdrawn 2007), https://www.astm.org/Standards/E214.htm. Accessed 19 Nov 2020

29. ASTM E1001-16, Standard Practice for Detection and Evaluation of Discontinuities by the Immersed Pulse-Echo Ultrasonic Method Using Longitudinal Waves, https://www.astm.org/Standards/E1001.htm. Accessed 19 Nov 2020
30. S. Elliot, *The Physics and Chemistry of Solids* (Wiley: Hoboken, NJ, USA, 1998), https://www.wiley.com/en-us/The+Physics+and+Chemistry+of+Solids-p-9780471981954. Accessed 19 Nov 2020
31. P.E. McHugh, Introduction to crystal plasticity theory, in *Mechanics of Microstructured Materials* (Springer: Cham, Switzerland, 2004), pp. 125–171
32. N. Pandech, K. Sarasamak, S. Limpijumnong, Elastic properties of perovskite A TiO_3 (A = Be, Mg, Ca, Sr, and Ba) and Pb B O_3 (B = Ti, Zr, and Hf): First principles calculations. J. Appl. Phys. **117**, 174108 (2015). https://doi.org/10.1063/1.4919837
33. H. Wang, P.L. Prendiville, P.J. McDonnell, W.V. Chang, An ultrasonic technique for the measurement of the elastic moduli of human cornea. J. Biomech. **29**, 1633–1636 (1996). https://doi.org/10.1016/S0021-9290(96)80017-0
34. D.E. Bray, R.K. Stanley, *Nondestructive Evaluation: A Tool in Design, Manufacturing and Service* (CRC Press: Boca Raton, FL, USA, 1996), https://books.google.com.et/books?id=5WtmjwEACAAJ&printsec=copyright#v=onepage&q&f=false. Accessed 19 Nov 2020
35. M. Rabiei, A. Palevicius, S. Nasiri, A. Dashti, A. Vilkauskas, G. Janusas, Relationship between young's modulus and planar density of unit cell, super cells (2 × 2 × 2), symmetry cells of perovskite ($CaTiO_3$) LATTICE. Materials **14**, 1258 (2021). https://doi.org/10.3390/ma14051258
36. M. Bodke, U. Gawai, A. Patil, B. Dole, Estimation of accurate size, lattice strain using Williamson-Hall models, SSP and TEM of Al doped ZnO nanocrystals. Mater. Tech. **106** (2018). https://doi.org/10.1051/mattech/2018055
37. R.S. Figliola, D.E. Beasley, *Theory and Design for Mechanical Measurements*, 7th ed. (Wiley: Hoboken, NJ, USA, 2014), https://www.wiley.com/en-us/Theory+and+Design+for+Mechanical+Measurements%2C+7th+Edition-p-9781119475651. Accessed 19 Nov 2020
38. K. Moradi, A.A. Sabbagh Alvani, First-principles study on Sr-doped hydroxyapatite as a biocompatible filler for photo-cured dental composites. J. Aust. Ceram. Soc. **56**, 591–598 (2020). https://doi.org/10.1007/s41779-019-00369-9
39. S.S. Bhat, U.V. Waghmare, U. Ramamurty, First-principles study of structure, vibrational, and elastic properties of stoichiometric and calcium-deficient hydroxyapatite. Cryst. Growth Des. **14**, 3131–3141 (2014). https://doi.org/10.1021/cg5004269
40. W.Y. Ching, P. Rulis, A. Misra, Ab initio elastic properties and tensile strength of crystalline hydroxyapatite. Acta Biomater. **5**, 3067–3075 (2009). https://doi.org/10.1016/j.actbio.2009.04.030
41. N.H. De Leeuw, J.R. Bowe, J.A.L. Rabone, A computational investigation of stoichiometric and calcium-deficient oxy- and hydroxy-apatites. Faraday Discuss. **134**, 195–214 (2007)
42. J.L. Katz, K. Ukraincik, On the anisotropic elastic properties of hydroxyapatite. J. Biomech. **4**, 221–227 (1971). https://doi.org/10.1016/0021-9290(71)90007-8
43. J.M. Zhang, Y. Zhang, K.W. Xu, V. Ji, Anisotropic elasticity in hexagonal crystals. Thin Solid Films **515**, 7020–7024 (2007). https://doi.org/10.1016/j.tsf.2007.01.045
44. V. Berdichevsky, Energy of dislocation networks. Int. J. Eng. Sci. **103**, 35–44 (2016). https://doi.org/10.1016/j.ijengsci.2016.02.011
45. R.E. Reed-Hill, R. Abbaschian, *Physical Metallurgy Principles* (Cengage: Boston, MA, USA, 1992)
46. D. Shi, G. Jiang, J. Bauer, The effect of structural characteristics on the in vitro bioactivity of hydroxyapatite. J. Biomed. Mater. Res. **63**, 71–78 (2002). https://doi.org/10.1002/jbm.10087
47. C.V. Ragel, M. Vallet-Regí, L.M. Rodríguez-Lorenzo, Preparation and in vitro bioactivity of hydroxyapatite/solgel glass biphasic material. Biomaterials **23**, 1865–1872 (2002). https://doi.org/10.1016/S0142-9612(01)00313-1
48. S. Dasgupta, S.S. Banerjee, A. Bandyopadhyay, S. Bose, Zn- and Mg-doped hydroxyapatite nanoparticles for controlled release of protein. Langmuir **26**, 4958–4964 (2010). https://doi.org/10.1021/la903617e

49. I. Uysal, F. Severcan, A. Tezcaner, Z. Evis, Co-Doping of hydroxyapatite with zinc and fluoride improves mechanical and biological properties of hydroxyapatite. Prog. Nat. Sci. Mater. Int. **24**, 340–349 (2014). https://doi.org/10.1016/j.pnsc.2014.06.004
50. P. Ptáček, *Apatites and their Synthetic Analogues—Synthesis, Structure, Properties and Applications* (InTech: London, UK, 2016)
51. C. Suryanarayana, M.G. Norton, C. Suryanarayana, M.G. Norton, Practical aspects of X-ray diffraction, in *X-ray Diffraction* (Springer: New York, NY, USA, 1998), pp. 63–94
52. K. Itatani, K. Tsuchiya, Y. Sakka, I.J. Davies, S. Koda, Superplastic deformation of hydroxyapatite ceramics with B_2O_3 or Na_2O addition fabricated by pulse current pressure sintering. J. Eur. Ceram. Soc. **31**, 2641–2648 (2011). https://doi.org/10.1016/j.jeurceramsoc.2011.01.014

Chapter 6
Novel Methodology for the Fabrication of In-Vitro Bioactive Scaffolds Comprised of Silver-Doped Hydroxyapatite Combined with Polyvinyltrimethoxysilane

This chapter presents the novel fabrication of an Ag-doped HA + PVTMS scaffold [1]. Recently, the focus has shifted to enhancing the biocompatibility and mechanical properties of biomaterials with highly porous structures. Porous composites represent a modern class of bioengineering materials with outstanding functional and structural attributes. This study explores the physical and mechanical characteristics of silver (Ag)-doped hydroxyapatite (HA) synthesized through mechanochemical and spark plasma sintering (SPS) methods. The study examines how the dopant affects phase formation, structural properties, mechanical properties, and morphological characteristics. Uniquely, a hair band was employed as a mold to produce a porous scaffold with an average size exceeding 100 μm. Employing the Monshi–Scherrer method, the scaffold's crystal size was determined to be 38 ± 2 nm, aligning well with the average values from transmission electron microscopy (TEM) analysis. The scaffold's compressive strength was tested, recording a peak value of ~ 15.71 MPa. Utilizing XRD, TEM, Fourier-transform infrared (FTIR), scanning electron microscope (SEM), and energy dispersive X-Ray analysis (EDAX), results affirmed the scaffold's bioactivity and indicated that both Ag-HA doping and polyvinyltrimethoxysilane (PVTMS) as an additive were beneficial. The findings highlighted that thermal treatment impacts the composition of Ag and HA, with no transformation changes occurring at 850 °C. Moreover, PVTMS significantly contributes as an additive by hindering decomposition and generating open microporosities in the scaffold, which enhance bioactivity.

6.1 Introduction

Considering the presence of Calcium Phosphates (CaPs) within the human body, researchers have explored the potential of CaPs for bone repair and replacement. Among the notable and widely recognized groups of CaPs is hydroxyapatite (HA)

M. Rabiei et al., *Hydroxyapatite-Based Nanocomposites*, Springer Series in Chemical Physics 127, https://doi.org/10.1007/978-3-032-12583-5_6

[2]. Its chemical composition is represented by Ca10(PO4)6(OH)2, which closely resembles bone tissue [3]. In crystallography, HA is known to have two structural forms: hexagonal and monoclinic. The hexagonal form is primarily preferred for biomedical applications due to its enhanced stability in bioactive conditions over the monoclinic form [4]. A notable challenge in employing HA as a bioactive agent is its susceptibility to infection, making the incorporation of antibacterial agents a crucial solution [5]. Studies have identified silver (Ag) as possessing significant antibacterial qualities [6–8]. Additionally, Ag stands out for its biocompatibility, boasting a complete antibacterial coefficient (100%) [9, 10]. Nevertheless, the compatibility of Ag is influenced by factors such as the strength, density, and final phase of the bioactive composites. Numerous mechanisms describe Ag's interaction with biological macromolecules [11]. Furthermore, Ag + ions are known to attach to protein functional groups, triggering protein denaturation [12]. Polyvinyltrimethoxysilane (PVTMS) features a silanol-like functional group (Si–O-H), which aids in bonding, making it effective for preventing composite degradation [13]. PVTMS, based on polysiloxane constituents, is particularly noteworthy due to its dense siloxane structure cross-linked with polymeric chains. Additionally, PVTMS is bioactive, improving mechanical attributes by establishing a robust Si–O-Si network [14]. Lately, the mechanochemical process has emerged as a novel strategy, wherein mixed powders undergo the impact of high-energy balls against the container, with the interaction persisting until a balance between bonding and fracture rates is achieved [15, 16]. Several factors affect the mechanochemical process, including grinding rate [16], grinding duration [17, 18], the ball-to-powder weight ratio [17], grinding atmosphere [17, 19], and the temperature applied [20]. Spark plasma sintering (SPS) is another method for sintering both conductive and non-conductive materials. Research on SPS began in 1930 [21]. The micro spark plasma is central to the general theory of SPS. This theory suggests that a localized electrical discharge leads to the rapid generation of a high-temperature spark plasma in a small area between powder particles within a second [22]. As previously stated, HA is highly sensitive to temperature, and the phase changes and decomposition caused by HA's high temperature, approximately 1100 °C, underpin the choice of the SPS process for sintering [23]. In this study, the use of SPS is particularly effective as it prevents collapse and inhibits grain growth, properties essential for enhancing bioactivity and mechanical properties [24]. Various studies have employed SPS for producing HA or HA-based composites. For instance, Shen et al. utilized the SPS technique to create transparent HA ceramics with a nano-grain structure at elevated temperatures [25]. Moreover, Zhang et al. explored the influence of HA on microstructure and biocomposites, using the SPS method for synthesis [26].

Simon et al. developed composites made of bioactive glass and polymer, employing poly-96L/4D lactide for enhanced strength [27]. Meanwhile, Goller et al. investigated the variables affecting the electrospinning of gelatin with silver nanoparticles and bioactive glass for applications in tissue engineering [28]. Nevertheless, previous studies incorporating silver for its antibacterial properties and polymer for structural reinforcement have not addressed the creation of porosity to boost biocompatibility, especially when using hydrophobic polymers such as those containing silane groups. To address the limitations related to pore size in composite scaffold fabrication, a novel method involving a hair band was suggested. This innovative multi-pore scaffold, featuring localized pores, was designed to enhance scaffolds' in vitro bioactivity response. The objective of this study was to develop a scaffold combining Ag-doped HA and PVTMS for the first time. The present work employed a mechanochemical process to synthesize Ag-doped HA. Based on the differential scanning calorimetry (DSC) results of the hair band soaked in slurry, the final sintering temperature was set at 350 °C to fabricate a porous scaffold. The resulting porosities were successfully opened to potentially enhance biocompatibility. The study examined the doping mechanisms, structural transformations, morphological characteristics, mechanical features, and bioactivity via spectroscopic and microscopic analyses. Notably, the innovative aspect of this research was using a hair band to generate porosity exceeding 100 μm and employing PVTMS to maintain the scaffold structure during heat treatment.

6.2 Experimental Methods

6.2.1 Materials and Instruments

In this study, calcium nitrate tetrahydrate ($Ca(NO_3)_2.4H_2O$), phosphorus pentoxide (P_2O_5), silver nitrate ($AgNO_3$) and *vinyltrimethoxysilane* (VTMS) (merck) were utilized as the precursors. Powders X-Ray, phase series were confirmed by X-ray diffraction (XRD) and performed on a Philips XRD diffractometer, Cukα radiation was used at 40 kV, 30 mA, step size of 0.05° (2Θ) nd scan rate of 1°/min (AXS GmbH, Kaunas, Lithuania). Furthermore, X'Pert software was used for qualitative analysis and report of width diffraction peaks (rad, β) at full width half maximum (*FWHM*) in different 2θ values according to the situation of peaks (Version 4.9.0). Moreover, differential scanning calorimetry (DSC) measurements were performed using STA-BAHR in Isfahan, Iran. Fourier-transform infrared spectroscopy (FTIR) spectra were obtained by embedding the compounds in potassium bromide (KBr) powders, utilizing a Perkin-Elmer Spectrum BX FT-IR spectrometer also located in Isfahan, Iran. Transmission electron microscopy (TEM) analysis was conducted with a Tecnai G2 F20 X–TWIN microscope operating at an acceleration voltage range of 50 to 80 kV in Tehran, Iran. To identify chemical elements within the components, an energy dispersive X-Ray (EDX) spectrometer Phillips/FEI 149 Quanta 200 was employed in Tehran, Iran. Scanning electron microscope (SEM) examination, using Phillips/FEI Quanta 200, was carried out to assess the morphology of the compounds and scaffold in Tehran, Iran. Additionally, simulated body fluid (SBF) was prepared following the protocol of Kokubo et al. [29]. At a constant 37 °C, materials were successively added and dissolved as per Kokubo's method. Then, $(CH_2OH)3CNH_2$ and HCl from Merck were added dropwise to adjust the pH to a final value of 7.40.

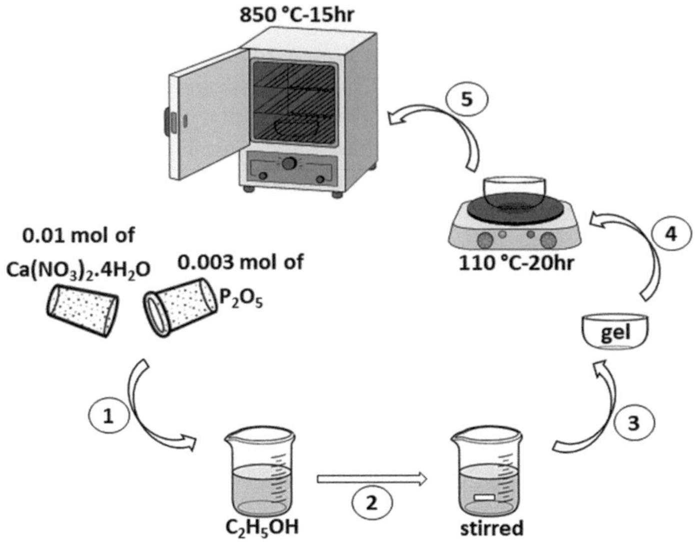

Fig. 6.1 The schematic flow diagram of the synthesis route of HA

Subsequently, the temperature was lowered to 20 °C, and distilled water was introduced [29]. In this research, the sample measurements were conducted in triplicate, and statistical results were presented as average values.

6.2.2 *Synthesis of HA*

According to Fig. 6.1, hydroxyapatite (HA) was synthesized by sol–gel method. Taking into account the Ca/P ratio of HA, calcium nitrate tetrahydrate ($Ca(NO_3)_2.4H_2O$) and phosphorus pentoxide (P_2O_5) were used in the molar ratio of 10:3. The following steps were performed: (1) $Ca(NO_3)_2.4H_2O$) and P_2O_5 were dissolved in 10 mL of ethyl alcohol (C_2H_5OH) and distillated water. (2) The product was stirred at 350 rpm for 2 h. (3) The gel was prepared at the bottom of the dish. 4) The gel was then dried at 110 °C in air for 20 h. 5) Heat treatment at 850 °C for 15 h was considered for sintering. This is similar to the method is presented in [30, 31]. The X-ray diffraction of synthesized HA is presented in Fig. 6.2.

6.2.3 *Polymerization of Vinyltrimethoxysilane*

Vinyltrimethoxysilane (VTMS) was polymerized to 20-mers. In addition, PVTMS was caused to enhance the thermal stability and physical properties of scaffold.

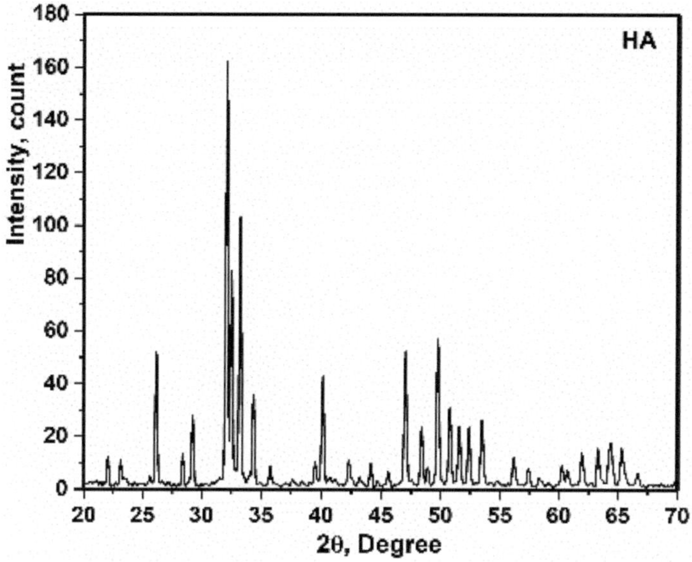

Fig. 6.2 X-ray diffraction of HA

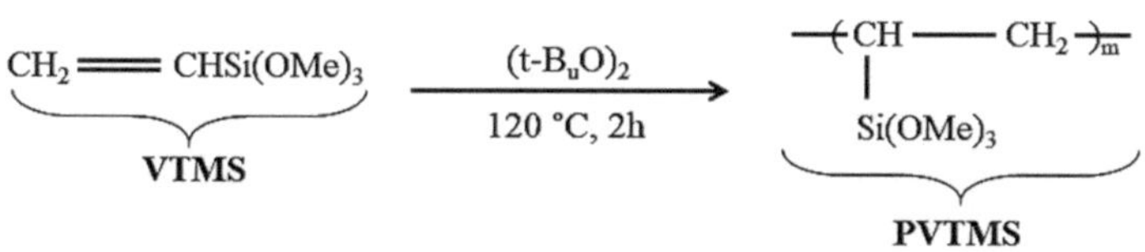

Fig. 6.3 Synthesis route of PVTMS

PVTMS was prepared with tertiary butyl peroxide as an initiator under reflux for 2 h at 150 °C in flowing nitrogen [32, 33]. In addition, the synthesis route of PVTMS is shown in Fig. 6.3.

6.2.4 Ag-Doped HA by Mechanochemical and Spark Plasma Sintering Process

In this study, silver nitrate ($AgNO_3$) with 99% purity was used as the dopant. The amount of replacement of Ca^{2+} ions by Ag^{+} corresponds to the formula $Ca_{10-X}Ag_X(PO_4)_6(OH)_{2-X}x$. In this case, X = 2 and x is represented as the x-mole of OH-vacancy formed by doping X-mole of Ag in the HA lattice. Moreover, the ratio of (Ca + Ag)/P was set to 1.67, while the degree of cationic substitution was changed. According to Fig. 6.4, (1) The powders of HA and $AgNO_3$ were ground by using a high-energy planetary ball mill via a mechanochemical process at 4 h.

In mechanochemical process, hard chromium steel shells and balls with a diameter of 20 mm were used. This process was carried out at ambient atmosphere and rate of 500 rpm with a ball to powder ratio of 15:1. In addition, to prevent agglomeration of the particles, the machine was stopped per 45 min. (2) After the end of mechanochemical process, calcination was carried out at 900 °C, at a heating rate of 10 °C per minute for degassing (90 min). (3) Taking into account the fact that HA is very sensitive to temperature, transformation and decomposition, the SPS process was considered. In this process, the powder was placed in the graphite mold (diameter = 20 mm), and pressure and temperature values were set 50 MPa and 600 °C for 10 min, respectively. The route of the fabricated Ag-doped HA composite is shown in Fig. 6.4.

6.2.5 New Approach of Fabrication Ag-Doped HA + PVTMS Scaffold

(4) Component of a slurry consisted of Ag-doped HA (93 wt%)/PVTMS (5 wt%)/ H_2O (1 wt%)/CH_3COOH (1 wt%) were prepared. (5) In addition, a hair band (commercial) was chosen as a new approach to obtain open porosities, (6) which was immersed in the slurry for 5 days at room temperature. According to Fig. 6.4, the hair band was cut and opened and the tape was pulled out, the taped hair band was immersed in the slurry and then the tape was rolled up. (7) Then, the product was placed in the furnace at 350 °C for 1 h according to the DSC curve, vinyl groups and temperature of the burning hair band. Finally, the porous scaffold consisted of Ag-doped HA + PVTMS was fabricated. It is necessary to mention that HA is considered as a bioactive matrix due to the presence of calcium phosphate groups. Moreover, Ag is the best metal for an antibacterial environment. In addition, acetic acid (CH_3COOH) has played an important role as a dispersant for particles, especially for Ag, to prevent the formation of colloidal particles and to prepare small particles. PVTMS is bioactive polymer, as well as PVTMS is an organo-silane molecule and provides a hydrophobic environment in the composite. Since water can easily diffuse into the polymer structure, and it can lead to the breakdown of intermolecular forces and create the voids in the polymer [34]. The effect of PVTMS is related to the implementation of the hydrophobic environment and the removal hydroxyl groups via silane groups in the PVTMS structure. Nevertheless, bonds of C and siloxane with hybrid sp^2, sp^3 are more stable and uniform, as well as, PVTMS is an intermediate for bonding and it will lead to improvement in mechanical properties and adhesion. Therefore, PVTMS prevents a hydrophilic ambient and the collapse of the scaffold [35].

PVTMS played an important role as a drying control chemical additive (DCCA) to prevent cracking and shrinkage of scaffold. As shown in Fig. 6.3, the free radicals of PVTMS structure can contribute to the flexibility of the scaffold through the bonding between C–C and siloxane (Si–O-Si). Furthermore, PVTMS has carbon chains, and

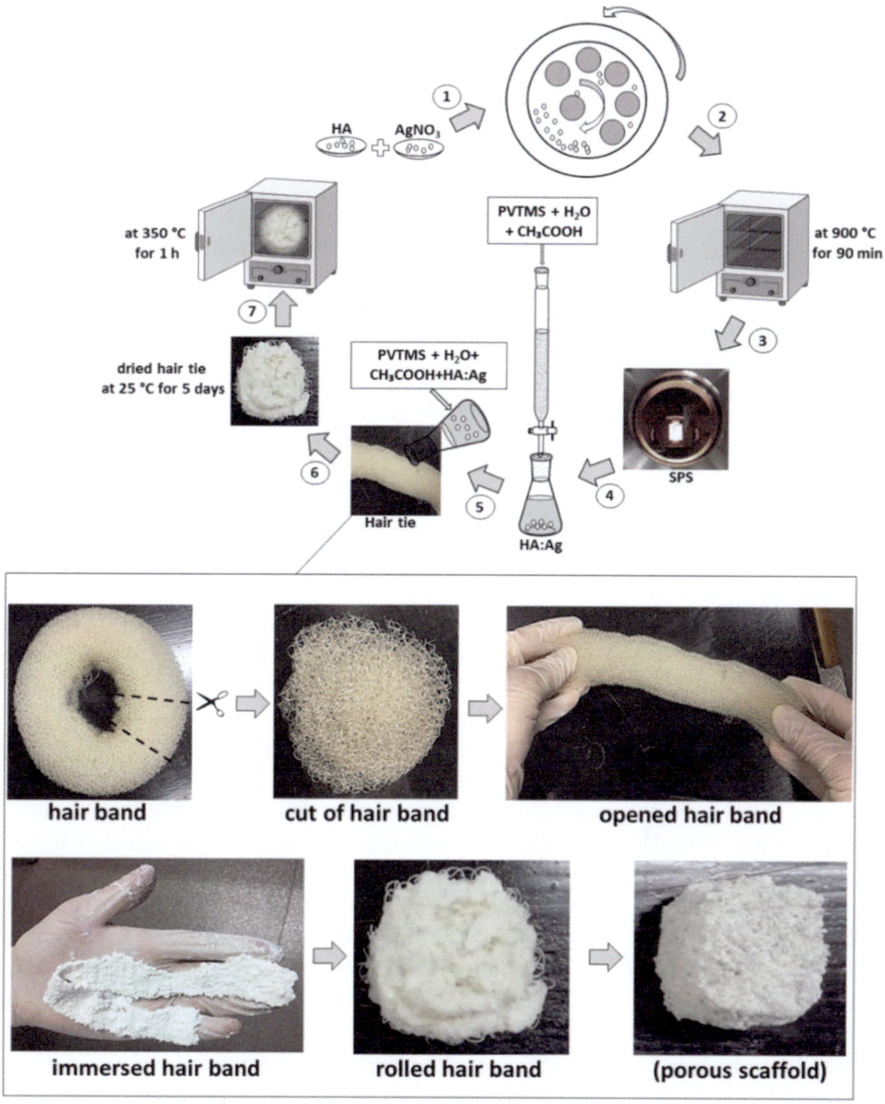

Fig. 6.4 The routes of fabricated Ag-doped HA + PVTMS scaffold

these chains can help to prevent the collapse of scaffold during the heat treatment, leaving solvent and volatile materials and creating porosity without damaging the scaffold.

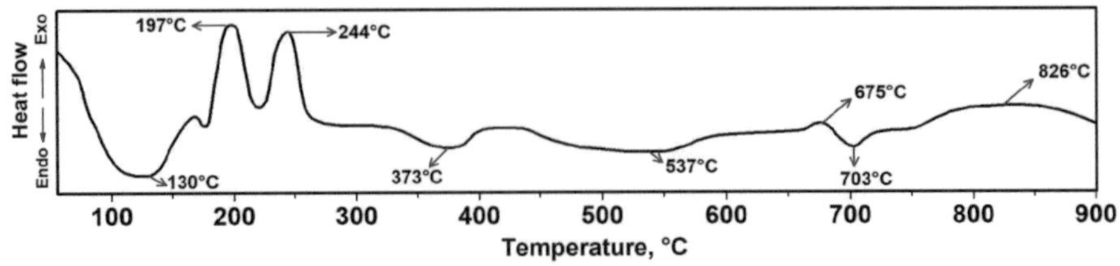

Fig. 6.5 DSC of rolled hair band consisted of Ag-doped HA + PVTMS + CH_3COOH + H_2O

6.3 Results and Discussion

6.3.1 DSC Analysis of Rolled Hair Band Consisted of Ag-Doped HA + PVTMS + CH_3COOH + H_2O

The DSC of the rolled hair band consisted of Ag-doped + HA + PVTMS + CH_3COOH + H_2O is shown in Fig. 6.5. The endothermic spectra at ~130 °C is related to the evaporation of physically absorbed water on the scaffold surface [36]. The exothermic peaks at 197 and 244 °C are related to the self-combustion process of the burning hair band with the evolution of volatile products [37]. Chemically absorbed water requires more energy to be released from the structure. The rapid decomposition of PVTMS occurs at 373 °C. Water in pores requires more heat to be released due to the capillary effect, so the final heat treatment temperature of 350 °C (<373 °C) was chosen with considering the collapse temperature. In addition, the peak at ~373 °C can be related to desorption of chemically bound absorbed lattice water, removal of O–H groups [30, 38] and at 537 and 675 °C are due to dehydration and at 703 °C it is due to condensation of HPO_4 and crystallization of HA and decarburization-phase transformation into HA such as beta-tricalcium phosphate (β-TCP) [39]. Accordingly, the exothermic temperature at ~ 832 °C can be attributed to the dehydroxylation of HA leading to the decomposition of HA to oxyhydroxyapatite [40, 41]. According to Fig. 6.5, the temperature to prevent degradation was chosen as 350 °C (for 1 h), and this temperature is an optimum value because at this temperature, the hair band is burned, and the porosities are created without damage and degradation to the scaffold. At high temperatures the composite is degraded, but we do not need high temperatures, because the body cannot tolerate more than normal temperature (> 39 °C); therefore, the thermal stability of 350 °C is perfectly suitable.

6.3.2 X-Ray Diffraction and TEM Analysis

X-ray diffraction of the Ag-doped HA + PVTMS scaffold is shown in Fig. 6.6. It is clear that the pattern is close to the pure HA pattern [3, 42]. The precursors such as

$Ca(NO_3)_2.4H_2O$), P_2O_5 and $AgNO_3$ disappear, and single phase of Ag-doped HA + PVTMS is produced. In addition, the absence of specific peaks of components such as $AgNO_3$ and Ag_3PO_4 and the agreement of the XRD pattern (Fig. 6.2) with the XRD pattern of pure HA were performed. Further, Ag^+ ions being replaced in the HA lattice are confirmed. Additionally, a slight shift of the peaks to low and high angles can be associated with the replacement of carbonate rather than phosphate groups. Moreover, the characteristic main peaks of (002) at $2\theta = 25.86°$, (211) at 31.86°, (300) at 32.89°, (130) 39.75°, (222) at 46.72°, (213) at 49.45° and (004) at 53.11° are investigated, showing that the cationic substitution not only prevents the formation of Ag-doped HA + PVTMS but also stabilizes at room temperature, as confirmed in [43]. Moreover, the peaks at 54.48°, 64.96° and 68.43° are attributed to the AgO and Ag tandemly [44]. Similarly, X-ray diffraction of Ag-doped HA is reported in Ref [45, 46]. The lattice parameters of the HA and Ag-doped HA + PVTMS structures were recorded as (a = 9.53 Å, c = 6.76 Å) and (a = 9.59 Å, c = 6.86 Å), respectively. Therefore, the a and c parameters of the Ag-doped HA + PVTMS structure increased, which is due to the larger ionic radius of Ag compared to Ca and it is related to the proper bonding between Ag ions and the HA structure [47].

According to the Monshi–Scherrer method, the nanocrystalline size of the Ag-doped HA + PVTMS scaffold was calculated. Furthermore, the linear plot of Ln β (β in radians) vs. Ln (1/(Cos Θ)) (degree) can be a linear plot for all chosen peaks (Fig. 6.7a). In Eq. 6.1, β is the full width at half maximum of the peak in radians, K is the shape factor, usually assumed to be 0.89 for ceramic materials, λ is the wavelength of the radiation in nanometers ($\lambda_{Cu_{K\alpha}} = 0.15405$ nm), Θ is the diffraction

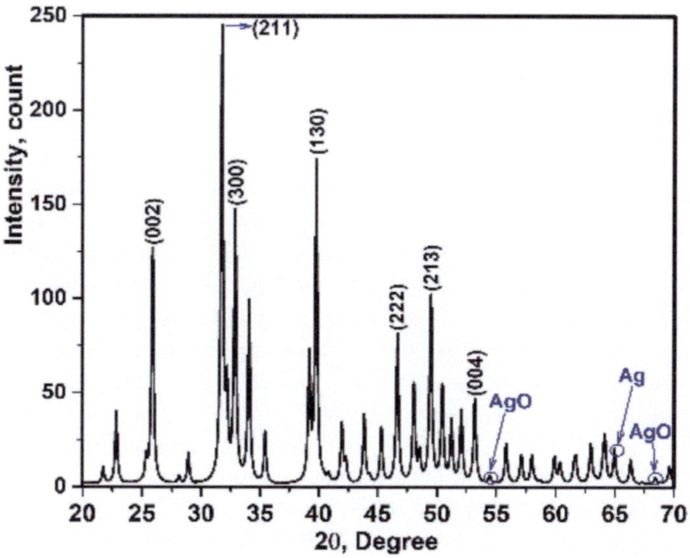

Fig. 6.6 X-ray diffraction of Ag-doped HA + PVTMS scaffold

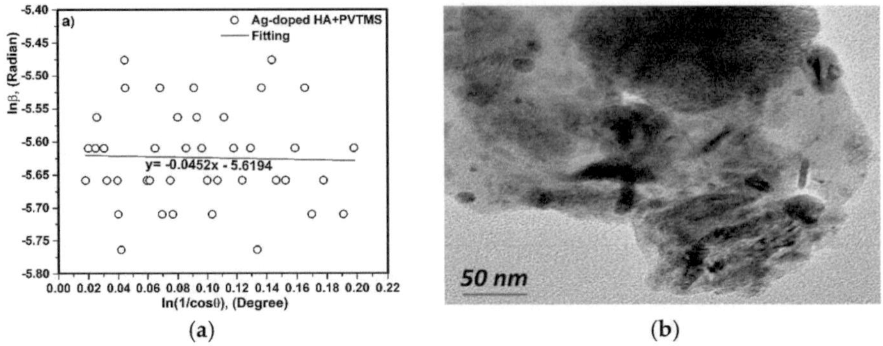

Fig. 6.7 **a** Linear plots of the Monshi-Scherrer equation and **b** TEM image of scaffold

angle of the peak and L is the nanocrystal size [3, 48]. The value of the intercept is calculated as $e^{(-5.6194)} = 0.0036$, therefore $\frac{K\lambda}{L} = 0.0036$, and the crystal size is calculated as $L = 38 \pm 2$ nm. In addition, the grain size was carried out on TEM image and (D_{TEM}) is ~ less than 50 nm (Fig. 6.7b). For scaffold, the D_{TEM} value is almost larger than the extracted crystallite size from Monshi–Scherrer equation, which can be explained by the fact that a grain consists of more than one crystallite and it is related to the nucleation and growth of the particles [48–51].

$$\mathrm{Ln}\,\beta = \mathrm{Ln}\left(\frac{K\lambda}{L}\right) + \mathrm{Ln}\left(\frac{1}{\mathrm{Cos}\,\theta}\right) \tag{6.1}$$

6.3.3 Fourier-Transform Infrared Spectroscopy (FTIR) Analysis

The FTIR spectrum of the Ag-doped HA + PVTMS scaffold is shown in Fig. 6.8. This spectrum corresponds with FTIR spectra of HA [52, 53]. Moreover, the strong stretching mode is associated with PO_4^{3-} at 1049 cm^{-1} and the broad spectrum is due to the bonding between O and H. The evidence of the existence of carbonate groups in HA corresponds to the wave numbers at 1442 and 1486 cm^{-1} (belonging to the $-CH_3$ group) and all $-CH_3$ groups have implied the existence of PVTMS. Moreover, the wave number values of 927 and 985 cm^{-1} are due to the effect of PVTMS [54]. Based on the study of Gibson et al. in the HA structure, the AB-type substitution of the phosphate by the carbonate can improve the bioactivity and mechanical properties as well as lowering the sintering temperature [55]. Therefore, according to the $AgNO_3$ content in the fabrication process of scaffold, there is no impressive tangible difference between the FTIR of pure HA and the FTIR of Ag-doped HA + PVTMS, except the intensity and broadening of the spectrum in some

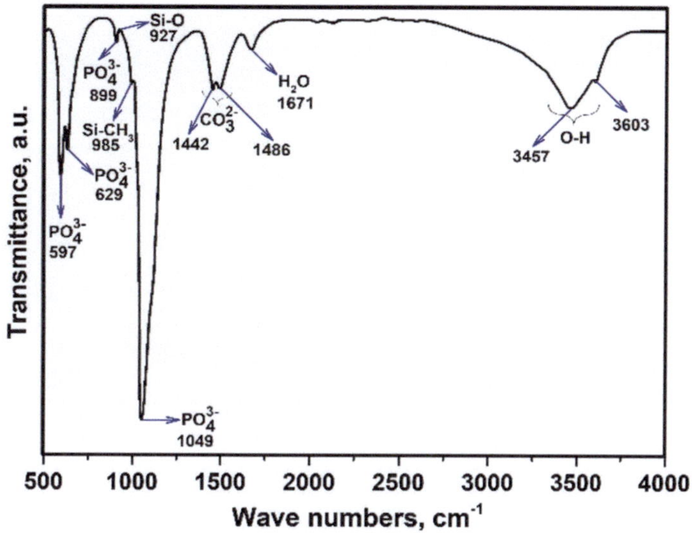

Fig. 6.8 FTIR spectra of Ag-doped HA + PVTMS scaffold

regions, which can be attributed to the substitutions of cations and anions in the HA structure. Furthermore the existence of main bonds are corresponded with [8, 10], overall, the absorption bands are characteristic bands for HA, which are the same as previously reported in [56].

6.3.4 Study of Morphology by SEM Analysis

The SEM image of Ag-doped HA powder is shown in Fig. 6.9. The background of HA has a disciplined distribution and can contribute to the properties of bioactivity [57]. It is clear that the Ag is eager to cover the surface of HA, and the growth of Ag grains is depicted. Meanwhile, some aggregate-like particles begin to appear on the HA crystals, which can be related to Ag_3PO_4 [58]. The chosen sintering temperature is greatly affected by the prevention of decomposition of Ag_3PO_4 into metallic Ag through sintering process.

In Fig. 6.10, SEM images of the Ag-doped HA + PVTMS scaffolds at various magnifications are presented, showing no signs of agglomeration. The heat treatment process caused O, H, and C to exit from the rolled hair bond, leading to the formation of porosities. These porosities, with sizes greater than 100 μm, contribute to a high bioactivity coefficient by facilitating the bonding between calcium and phosphate groups. The images indicate a uniform porosity distribution with an average size of approximately 200 μm, which is ideal for the migration of osteoblasts [59]. A significant achievement of this study is the creation of large porosities without compromising the scaffold's integrity. Prior research suggests that for effective bone growth,

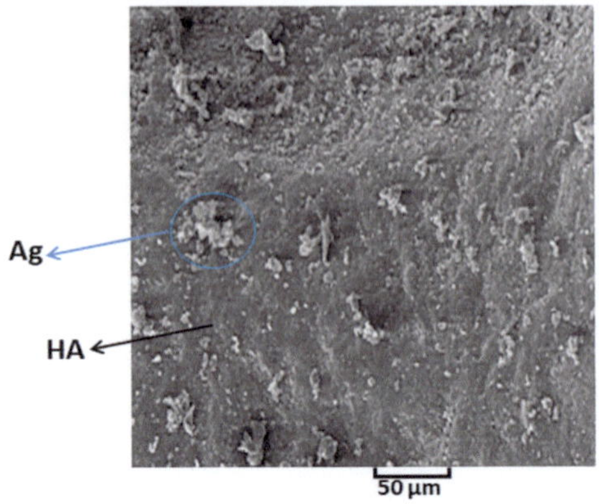

Fig. 6.9 SEM image of Ag-doped HA powder

the minimum porosity should be between 75–100 μm, with an optimal range of 100–135 μm [60, 61]. This study produces even larger porosities, exceeding 200 μm. Nonetheless, understanding the upper limits of pore size is crucial, as increased pore volume may affect the mechanical properties of scaffolds with larger pores [62].

Figure 6.11 presents the outcomes of an elemental analysis performed using EDAX stoichiometry in conjunction with SEM. The EDAX stoichiometry values indicate that, barring Cu—used here as a benchmark—no impurity elements are detected. The Ag-doped HA + PVTMS scaffold synthesized in this study exhibits a Ca/P ratio of 1.79, which closely matches the established Ca/P ratio of pure hydroxyapatite, reported as 1.67 [63]. The scaffold's minimal Ag content results from its initial AgNO3 composition, with the silver primarily contributing to its antibacterial properties, as the amount used is sufficient to support this characteristic. Notably, the substitution of Ag for Ca is critical, influencing hydroxyapatite's chemical characteristics, reducing its crystallinity, and enhancing its solubility [64]. The EDAX analysis confirms the presence of all expected constituent elements within the Ag-doped HA + PVTMS scaffold under study.

6.3.5 Mechanical Properties

Figure 6.12 illustrates the stress–strain compression curve for the Ag-doped HA + PVTMS scaffold. The compressive strength peaks at 15.71 MPa when the strain reaches approximately 0.77. Applying Hooke's law and choosing two distinct points within the elastic region yields an elastic coefficient of about 3.86 MPa. This value is lower than that of standard natural HA, which is attributed to the scaffold's porosity; the larger porosity size prioritizes bioactivity, as noted in [2, 65]. Additionally, the maximum compressive strength value was not accurately forecasted when using a hair band as a mold. This value could be suitable because, 1) fabrication of scaffold through the using components such as Ag-doped HA (93 wt%)/PVTMS (5 wt%)/

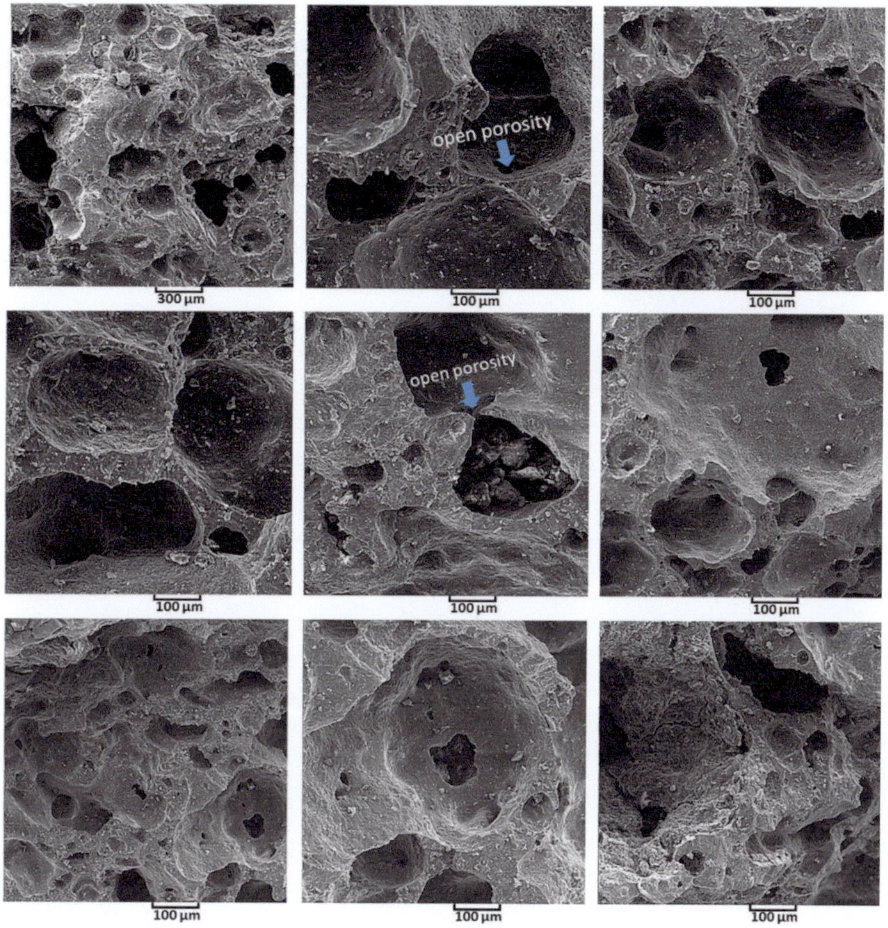

Fig. 6.10 SEM images of Ag-doped HA + PVTMS scaffold

H_2O (1 wt%)/CH_3COOH (1 wt%) without using metals with high concentration. 2) In this case, the bioactivity characteristics and creating the large size of porosities were purpose and, considering the initial bioactivity of ingredients, use of high concentration metals such as Mg, Zn and Pt can improve the value of compressive strength and mechanical properties.

6.3.6 Investigation of Bioactivity

Scaffold was immersed in SBF (Fig. 6.13) and kept in the oven at 37 °C (similar to body temperature). After 3, 5, 10 and 20 days, the scaffold was pulled out and

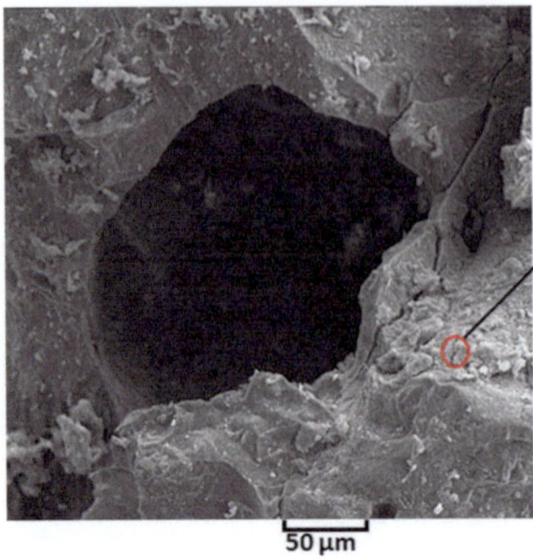

Element	Weight %
C	27.03
O	44.08
Ca	16.13
P	8.99
Ag	1.73
Si	1.03
Cu	1.01
Total = 100	

Fig. 6.11 SEM EDAX data for Ag-doped HA + PVTMS scaffold

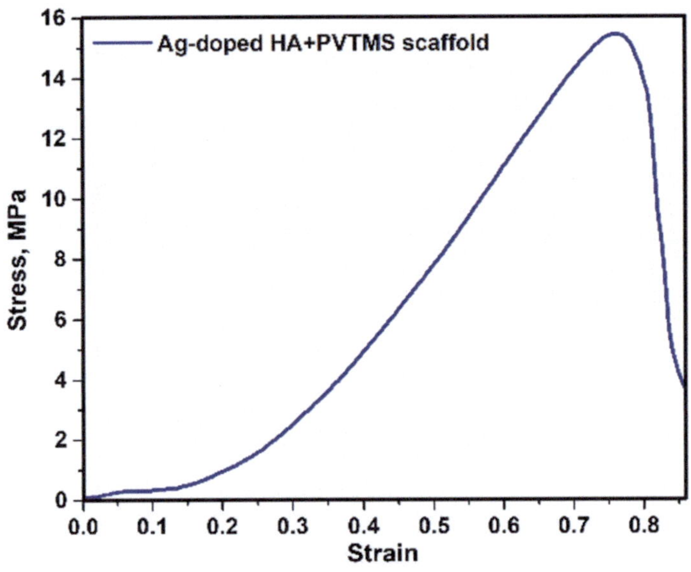

Fig. 6.12 The curve of stress–strain compression of Ag-doped HA + PVTMS scaffold

washed with distilled water then, analysis such as XRD, FTIR, SEM and EDAX were performed.

X-ray diffraction and FTIR spectra, along with SEM images and EDAX data analysis of the scaffold post-immersion in SBF, are displayed in Fi6.14–6.16. The analysis confirms no additional elements or phases are present. Interestingly, no imperfections in crystallization or ion substitutions were observed, aligning with the sharp peaks (indicating no amorphous content), and Ag fully covered the HA surface

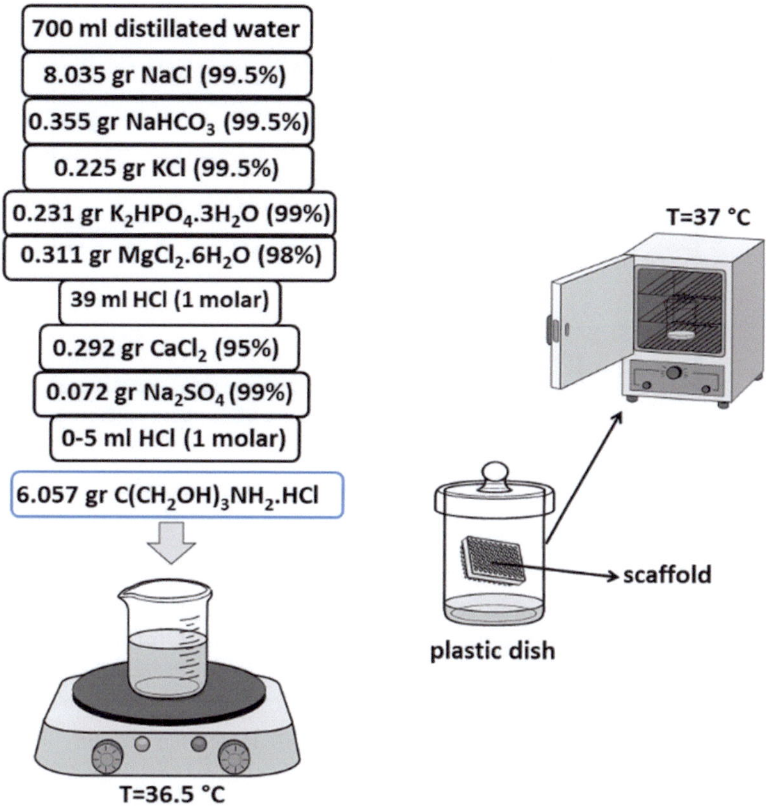

Fig. 6.13 Synthesis of SBF [29], immersed scaffold in the SBF at 37 °C

without notable substitutions. It is evident that after 3 days, the bonding between calcium and phosphate enhanced, and the X-ray diffraction reveals the HA pattern (reflecting the HA matrix) as seen in Fig. 6.14. Additionally, after 20 days, peak intensity increased, and the pattern indicates that the strong diffraction peaks at 2Θ values correspond to the HA structure, with hkl values of natural HA peaks assigned to 002, 102, 210, 211, 112, 300, 202, 310, 222, and 213, respectively [66]. A key finding of this analysis is the evident presence of Ag and PVTMS within the scaffold and significant porosity, which facilitates the nucleation and growth of calcium and phosphorus ions from SBF [67], resulting in a rapid increase in bioactivity seen with the formation of HA after just 3 days of immersion. Consequently, the selection of a porous hair band form proved highly effective.

In the FTIR spectrum (Fig. 6.15), the weak peaks at ~1600–2000 cm^{-1} can be associated with the Si–O-Si groups originating from PVTMS or with the overlap of spectra attributed to the carbonate and silanol groups [68]. Moreover, the weak peak at 1766 cm^{-1} can be attributed to the potential occurrence of hydrogen bonding between the carbonyl, hydroxyl and phosphate groups. This evidence indicates the

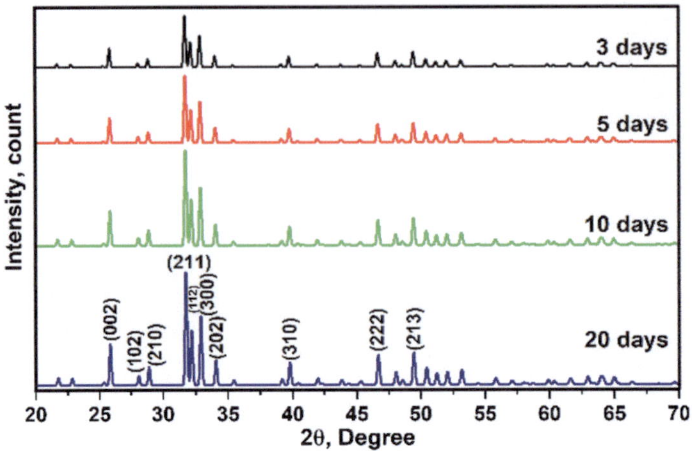

Fig. 6.14 X-ray diffraction of immersed scaffold in SBF after 3, 5, 10 and 20 days

presence of a bioactive scaffold in the structure. The transmittance values increased over the days, suggesting more precipitation of HA on the surface of the porous scaffold. In addition, a broad peak in the range from 2800 to 3760 cm^{-1} is associated with the bonding between O and H, which occurs as a bending mode, and the reason for the presence of these peaks is related to the hydroxyl groups of the precursor and water. Additionally, the researchers reported that the peak at 1480 cm^{-1} belongs to C–O [69, 70]. Thus, the FTIR spectrum agreed with the FTIR spectrum of natural HA [71].

Based on the SEM images presented in Fig. 6.16, the fabrication of the viscose slurry, integration of porosities, and incorporation of a hair band as an innovative mold were beneficial. The scaffold's surface saw a rapid rate of HA precipitation, with porosities being completely filled just 3 days after immersion (Fig. 6.16). This hair band molding technique enhanced the diffusion of calcium and phosphorus ions from the SBF into the porosities. Over time, the typical needle-like HA morphology [72] transitioned to spherical forms. As noted in references [73] and [74], the porosity sizes cater to the dimensions of blood cells, facilitating blood cell accommodation and enhancing bioactivity. Regarding mechanical properties, an increase in both porosity size and quantity correlates with enhanced bioactivity [75]. Figure 6.16 illustrates the upward trend of HA precipitation as immersion time extends. In terms of HA precipitation mechanics, HCl present in SBF boosts this process by heightening polar group concentrations on the scaffold surface, consequently increasing absorption rates [76, 77]. Figure 6.16 also details the EDAX results showing the HA presence over durations of 3, 5, 10, and 20 days. The scaffold's Ca/P ratios during immersion in SBF stood at 1.71, 2.07, 1.53, and 1.64, respectively. After 20 days, the Ca/P ratio of 1.64 approached the natural HA ratio of 1.67 [78]. Our research suggests that both in vivo tests and the development of scaffolds incorporating a bioactive metal component to enhance mechanical properties remain areas open for exploration.

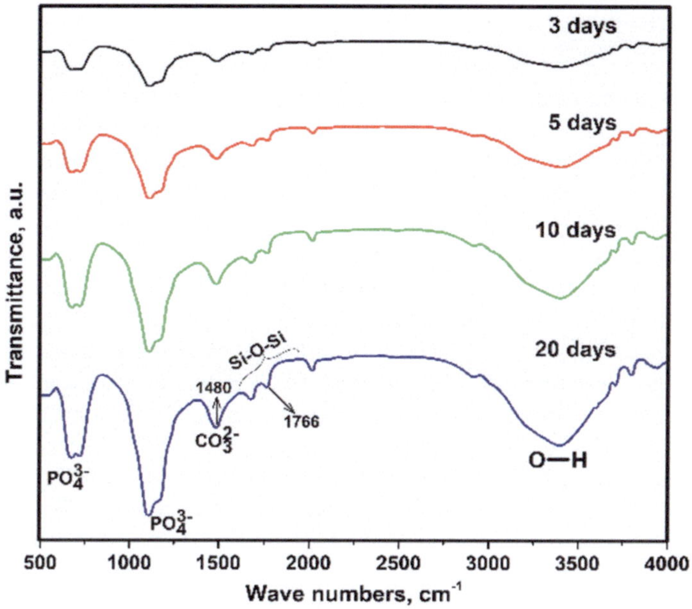

Fig. 6.15 FTIR spectrum of immersed scaffold in SBF after 3, 5, 10 and 20 days

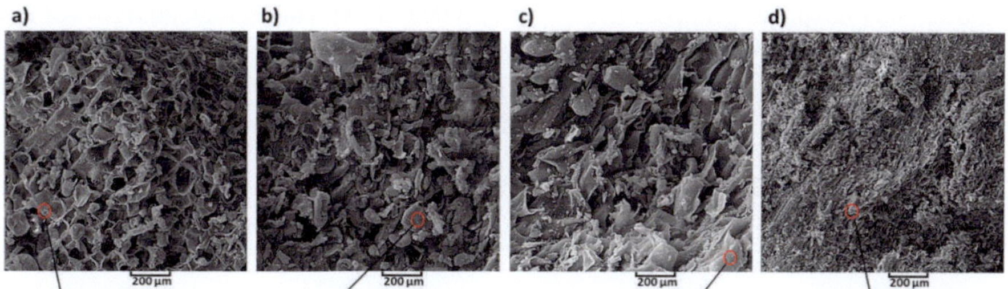

Element	a) Weight %	b) Weight %	c) Weight %	d) Weight %
C	26.11	21.09	19.11	21.04
O	39.29	47.41	41.44	34.19
Ca	19.18	18.99	21.48	26.15
P	11.21	9.14	13.97	15.92
Ag	1.29	1.01	0.89	0.82
Si	0.83	0.79	0.73	0.69
Cu	2.09	1.57	2.38	1.19
Total = 100	Total = 100	Total = 100	Total = 100	Total = 100

Fig. 6.16 SEM EDAX data of immersed scaffold in SBF after **a** 3 days, **b** 5 days, **c** 10 days and **d** 20 days

6.4 Conclusions

The creation of the composite involved the successful mechanochemical synthesis of Ag-doped HA. The SPS process notably prevented both temperature fluctuations and phase decomposition of the composite, while also ensuring uniform fabrication of Ag-doped HA. Consequently, the SPS process facilitated a desirable distribution of Ag within HA. Additionally, the impact of Ag and PVTMS loading on HA was evaluated. Ag created an antibacterial environment, while PVTMS inhibited scaffold cracking and shrinkage. The free radicals in the PVTMS structure enhanced scaffold flexibility through C–C and siloxane (Si–O-Si) bond formation, and its hydrophobic nature due to the silane group prevented collapse via carbon chains during heat treatment. The crystallite size of Ag-doped HA was determined to be 38 $\pm$ 2 nm, aligning with TEM analysis findings (~less than 50 nm). The a and c lattice parameters ($a = 9.59$ Å, $c = 6.86$ Å) increased in the Ag-doped HA + PVTMS structure because of Ag's larger ionic radius compared to Ca, facilitating strong bonding between Ag ions and the HA structure. A novel method employing a hair band was used to fabricate a porous scaffold, yielding an average porosity value greater than 200 μm, which accommodated blood cells effectively and enhanced bioactivity. For in vitro bioactivity testing, an SBF solution was synthesized successfully, and the scaffold was immersed in SBF for periods of 3, 5, 10, and 20 days. As evidenced by the FTIR spectrum, the immersed scaffold showed bands attributed to HA, predominantly linked to its phosphate groups. Furthermore, XRD, SEM, and EDAX analyses of the immersed scaffold revealed HA nucleation and growth on its surface, confirming its bioactivity. The maximum observed compressive strength was 15.71 MPa, deemed appropriate given the Ag content and minimal metal usage.

References

1. M. Rabiei, A. Palevicius, R. Ebrahimi-Kahrizsangi, S. Nasiri, A. Vilkauskas, G. Janusas, New approach for preparing In vitro bioactive scaffold consisted of Ag-doped hydroxyapatite + polyvinyltrimethoxysilane. Polymers. Basel: MDPI. **13**(11), 1–19, art. no. 1695 (2021). ISSN 2073-4360. https://doi.org/10.3390/polym13111695. https://www.mdpi.com/2073-4360/13/11/1695
2. R. Marzieh, N. Sohrab, P. Arvydas, J. Giedrius, Preparation and investigation of bioactive organic-inorganic nano-composite derived from PVB-co-VA-co-VAc/HA, in *Proceedings of the 15th International Conference Mechatronic Systems and Materials, MSM*, Bialystok, Poland, Institute of Electrical and Electronics Engineers Inc.: Piscataway, NJ, USA (2020)
3. M. Rabiei, A. Palevicius, A. Monshi, S. Nasiri, A. Vilkauskas, G. Janusas, Comparing methods for calculating nano crystal size of natural hydroxyapatite using X-ray diffraction. Nanomaterials **10**, 1627 (2020). https://doi.org/10.3390/nano10091627
4. K. Sato, Mechanism of hydroxyapatite mineralization in biological systems. J. Ceram. Soc. Jpn. **115**, 124–130 (2007). https://doi.org/10.2109/jcersj.115.124
5. V.K. Sharma, R.A. Yngard, Y. Lin, Silver nanoparticles: green synthesis and their antimicrobial activities. Adv. Colloid Interface Sci. **145**, 83–96 (2009)

6. J.S. Kim, E. Kuk, K.N. Yu, J.H. Kim, S.J. Park, H.J. Lee, S.H. Kim, Y.K. Park, Y.H. Park, C.Y. Hwang et al., Antimicrobial effects of silver nanoparticles. Nanomed. Nanotechnol. Biol. Med. **3**, 95–101 (2007). https://doi.org/10.1016/j.nano.2006.12.001
7. M. Mardziah, S. Iis S. Ramesh, Strontium-doped hydroxyapatite nanopowder via sol-gel method: effect of strontium concentration and calcination temperature on phase behavior. Trends Biomater. Artif. Organs**23**, 105–113. Accessed 16 Apr 2021. (2009). https://www.researchgate.net/publication/249011460_Strontium-Doped_Hydroxyapatite_Nanopowder_via_Sol-Gel_Method_Effect_of_Strontium_Concentration_and_Calcination_Temperature_on_Phase_Behavior
8. R.G. Saratale, G.D. Saratale, G. Ghodake, S.K. Cho, A. Kadam, G. Kumar, B.H. Jeon, D. Pant, A. Bhatnagar, H.S. Shin, Wheat straw extracted lignin in silver nanoparticles synthesis: Expanding its prophecy towards antineoplastic potency and hydrogen peroxide sensing ability. Int. J. Biol. Macromol. **128**, 391–400 (2019). https://doi.org/10.1016/j.ijbiomac.2019.01.120
9. L. Badrour, A. Sadel, M. Zahir, L. Kimakh, A. El Hajbi, Synthesis and physical and chemical characterization of Ca10-xAgx(PO4)6(OH) 2–xx apatites. Ann. Chim. Sci. Des Mater. **23**, 61–64 (1998). https://doi.org/10.1016/S0151-9107(98)80012-3
10. G.D. Saratale, R.G. Saratale, D.S. Kim, D.Y. Kim, H.S. Shin, Exploiting fruit waste grape pomace for silver nanoparticles synthesis, assessing their antioxidant, antidiabetic potential and antibacterial activity against human pathogens: a novel approach. Nanomaterials **10**, 1–18 (2020). https://doi.org/10.3390/nano10081457
11. C.S. Ciobanu, S.L. Iconaru, C. Chiiuc, A. Costescu, P. Le Coustumer, D. Predoi, Synthesis and antimicrobial activity of silver-doped hydroxyapatite nanoparticles. Biomed Res. Int.**2013** (2013). https://doi.org/10.1155/2013/916218
12. S.K. Rastogi, V.J. Rutledge, C.L. Gibson, D.A. Newcombe, J.R. Branen, A.L. Branen, Ag colloids and Ag clusters over EDAPTMS-coated silica nanoparticles: synthesis, characterization, and antibacterial activity against Escherichia coli. Nanomed. Nanotechnol. Biol. Med. **7**, 305–314 (2011). https://doi.org/10.1016/j.nano.2010.11.003
13. M. Lim, D. Kim, J. Seo, H. Han, Preparation and properties of poly(vinyl alcohol)/Vinyltrimethoxysilane (PVA/VTMS) hybrid films with enhanced thermal stability and oxygen barrier properties. Macromol. Res. **22**, 1096–1103 (2014). https://doi.org/10.1007/s13233-014-2146-2
14. A. Motalebi, M. Nasr-Esfahani, R. Ali, M. Pourriahi, Improvement of corrosion performance of 316L stainless steel via PVTMS/henna thin film. Prog. Nat. Sci. Mater. Int. **22**, 392–400 (2012). https://doi.org/10.1016/j.pnsc.2012.10.006
15. S.H. Rhee, Synthesis of hydroxyapatite via mechanochemical treatment. Biomaterials **23**, 1147–1152 (2002). https://doi.org/10.1016/S0142-9612(01)00229-0
16. C. Suryanarayana, *Mechanical Alloying and Milling; CRC Press: Boca Raton* (FL, USA, 2004)
17. P. Baláž, M. Achimovicová, M. Baláž, P. Billik, C.Z. Zara, J.M. Criado, F. Delogu, E. Dutková, E. Gaffet, F.J. Gotor et al., Hallmarks of mechanochemistry: From nanoparticles to technology. Chem. Soc. Rev. **42**, 7571–7637 (2013). https://doi.org/10.1039/c3cs35468g
18. V.N. Ermakov, S. Kruchinin, Electronic nanosensors based on nanotransistor with bistability behaviour. in *electron transport in nanosystems*, eds. by J. Bonča, S. Kruchinin, pp. 341–349 (Springer: Berlin/Heidelberg, Germany, 2008). https://doi.org/10.1007/978-1-4020-9146-9_26
19. S. Kruchinin, T. Pruschke, Thermopower for a molecule with vibrational degrees of freedom. Phys. Lett. Sect. A Gen. At. Solid State Phys.**378**, 1157–1161 (2014). https://doi.org/10.1016/j.physleta.2014.01.054
20. P. Baláž, Mechanochemistry in Nanoscience and Minerals Engineering (Springer: Berlin/Heidelberg, Germany, 2008). ISBN 9783540748540
21. O. Guillon, J. Gonzalez-Julian, B. Dargatz, T. Kessel, G. Schierning, J. Räthel, M. Herrmann, Field-assisted sintering technology/spark plasma sintering: Mechanisms, materials, and technology developments. Adv. Eng. Mater. **16**, 830–849 (2014). https://doi.org/10.1002/adem.201300409

22. Z.A. Munir, U. Anselmi-Tamburini, M. Ohyanagi, The effect of electric field and pressure on the synthesis and consolidation of materials: A review of the spark plasma sintering method. J. Mater. Sci. **41**, 763–777 (2006)
23. V.N. Ermakov, S.P. Kruchinin, T. Pruschke, J.K. Freericks, Thermoelectricity in tunneling nanostructures. Phys. Rev. B Condens. Matter Mater. Phys.**92**, 155431 (2015). https://doi.org/10.1103/PhysRevB.92.155431
24. T. Vanherck, G. Jean, M. Gonon, J. Lobry, F. Cambier, Spark plasma sintering: Homogenization of the compact temperature field for non conductive materials. Int. J. Appl. Ceram. Technol. **12**, E1–E12 (2015). https://doi.org/10.1111/ijac.12187
25. M. Eriksson, Y. Liu, J. Hu, L. Gao, M. Nygren, Z. Shen, Transparent hydroxyapatite ceramics with nanograin structure prepared by high pressure spark plasma sintering at the minimized sintering temperature. J. Eur. Ceram. Soc. **31**, 1533–1540 (2011). https://doi.org/10.1016/j.jeurceramsoc.2011.03.021
26. Y.H. He, Y.Q. Zhang, Y.H. Jiang, R. Zhou, Effect of HA (Hydroxyapatite) content on the microstructure, mechanical and corrosion properties of ([Formula presented])-xHA biocomposites synthesized by sparkle plasma sintering. Vacuum **131**, 176–180 (2016). https://doi.org/10.1016/j.vacuum.2016.06.015
27. A. Vulpoi, V. Simon, H. Ylänen, S. Simon, Development and in vitro assessment of bioactive glass/polymer nanostructured composites with silver. J. Compos. Mater. **48**, 63–70 (2014). https://doi.org/10.1177/0021998312468186
28. A. Akturk, M. Erol Taygun, G. Goller, Optimization of the electrospinning process variables for gelatin/silver nanoparticles/bioactive glass nanocomposites for bone tissue engineering. Polym. Compos. **41**, 2411–2425 (2020). https://doi.org/10.1002/pc.25545
29. T. Kokubo, H. Takadama, How useful is SBF in predicting in vivo bone bioactivity? Biomaterials **27**, 2907–2915 (2006). https://doi.org/10.1016/j.biomaterials.2006.01.017
30. K. Agrawal, G. Singh, D. Puri, S. Prakash, Synthesis and Characterization of hydroxyapatite powder by sol-gel method for biomedical application. J. Miner. Mater. Charact. Eng. **10**, 727–734 (2011). https://doi.org/10.4236/jmmce.2011.108057
31. I.S. Kim, P.N. Kumta, Sol-gel synthesis and characterization of nanostructured hydroxyapatite powder. Mater. Sci. Eng. B Solid-State Mater. Adv. Technol.**111**, 232–236 (2004). https://doi.org/10.1016/j.mseb.2004.04.011
32. Y. Abe, T. Namiki, K. Tuchida, Y. Nagao, T. Misono, Preparation and properties of silicon-containing hybrid gels from vinyltrimethoxysilane. J. Non Cryst. Solids **147–148**, 47–51 (1992). https://doi.org/10.1016/S0022-3093(05)80591-5
33. K. Tsuru, S. Hayakawa, C. Ohtsuki, A. Osaka, Bioactive gel coatings derived from vinyltrimethoxysilane. J. Sol-Gel Sci. Technol. **12**, 237–240 (1998)
34. H.G.P. Lewis, T.B. Casserly, K.K. Gleason, Hot-filament chemical vapor deposition of organosilicon thin films from hexamethylcyclotrisiloxane and octamethylcyclotetrasiloxane. J. Electrochem. Soc. **148**, F212 (2001). https://doi.org/10.1149/1.1415723
35. B. Arkles, Hydrophobicity, hydrophilicity and silane surface modification. Paint Coat. Ind. **22**, 114 (2006)
36. Y.Y. Özbek, F.E. Baştan, F. Üstel, Synthesis and characterization of strontium-doped hydroxyapatite for biomedical applications. J. Therm. Anal. Calorim. **125**, 745–750 (2016). https://doi.org/10.1007/s10973-016-5607-3
37. M. Streckova, E. Mudra, M. Sebek, T. Sopcak, J. Dusza, J. Kovac, Preparation and investigations of Ni0:2Zn0:8Fe2O4 ferrite nanofiber membranes by needleless electrospinning method. Acta Phys. Pol. A **131**, 729–731 (2017)
38. F. Miyaji, Y. Kono, Y. Suyama, Formation and structure of zinc-substituted calcium hydroxyapatite. Mater. Res. Bull. **40**, 209–220 (2005). https://doi.org/10.1016/j.materresbull.2004.10.020
39. C.S. Chai, B. Ben-Nissan, Bioactive nanocrystalline sol-gel hydroxyapatite coatings. J. Mater. Sci. Mater. Med. **10**, 465–469 (1999). https://doi.org/10.1023/A:1008992807888
40. C.J. Tredwin, A.M. Young, E.A. Abou Neel, G. Georgiou, J.C. Knowles, Hydroxyapatite, fluor-hydroxyapatite and fluorapatite produced via the sol-gel method: Dissolution behaviour

and biological properties after crystallisation. J. Mater. Sci. Mater. Med. **25**, 47–53 (2014). https://doi.org/10.1007/s10856-013-5050-y
41. K. Tõnsuaadu, K.A. Gross, L. Pluduma, M. Veiderma, A review on the thermal stability of calcium apatites. J. Therm. Anal. Calorim. **110**, 647–659 (2012)
42. A. Chandrasekar, S. Sagadevanand, A. Dakshnamoorthy, Synthesis and characterization of nano-hydroxyapatite (n-HAP) using the wet chemical technique. Int. J. Phys. Sci. **8**, 1639–1645 (2013). https://doi.org/10.5897/IJPS2013.3990
43. B. Singh, A.K. Dubey, S. Kumar, N. Saha, B. Basu, R. Gupta, In vitro biocompatibility and antimicrobial activity of wet chemically prepared Ca10-xAgx(PO4)6(OH) 2 (0.0 $\leq$ x $\leq$ 0.5) hydroxyapatites. Mater. Sci. Eng. C**31**, 1320–1329 (2011). https://doi.org/10.1016/j.msec.2011.04.015
44. A. Dubnika, D. Loca, V. Rudovica, M.B. Parekh, L. Berzina-Cimdina, Functionalized silver doped hydroxyapatite scaffolds for controlled simultaneous silver ion and drug delivery. Ceram. Int. **43**, 3698–3705 (2017). https://doi.org/10.1016/j.ceramint.2016.11.214
45. R.J. Chung, M.F. Hsieh, C.W. Huang, L.H. Perng, H.W. Wen, T.S. Chin, Antimicrobial effects and human gingival biocompatibility of hydroxyapatite sol-gel coatings. J. Biomed. Mater. Res. Part B Appl. Biomater**76**, 169–178 (2006). https://doi.org/10.1002/jbm.b.30365
46. L. Yang, X. Ning, Q. Xiao, K. Chen, H. Zhou, Development and characterization of porous silver-incorporated hydroxyapatite ceramic for separation and elimination of microorganisms. J. Biomed. Mater. Res. B Appl. Biomater**81**, 50–56 (2007). https://doi.org/10.1002/jbm.b.30635
47. N. Kose, A. Otuzbir, C. Pekşen, A. Kiremitçi, A. Doğan, A silver ion-doped calcium phosphate-based ceramic nanopowder-coated prosthesis increased infection resistance. Clin. Orthop. Relat. Res. **471**, 2532–2539 (2013). https://doi.org/10.1007/s11999-013-2894-x
48. M. Rabiei, A. Palevicius, S. Nasiri, A. Dashti, A. Vilkauskas, G. Janusas, Relationship between young's modulus and planar density of unit cell, super cells (2 $\times$ 2 $\times$ 2), symmetry cells of perovskite (CaTiO3) lattice. Materials **14**, 1258 (2021). https://doi.org/10.3390/ma14051258
49. E.M. Benali, A. Benali, M. Bejar, E. Dhahri, M.P.F. Graca, M.A. Valente, P. Sanguino, B.F.O. Costa, Effect of annealing temperature on structural, morphological and dielectric properties of La 0.8 Ba 0.1 Ce 0.1 FeO 3 perovskite. J. Mater. Sci. Mater. Electron**31,** 16220–16234 (2020). https://doi.org/10.1007/s10854-020-04140-w
50. M. Hosseinnezhad, J. Movahedi, S. Nasiri, High stability photosensitizers for dye-sensitized solar cells: Synthesis, characterization and optical performance. Opt. Mater. **109**, 110198 (2020). https://doi.org/10.1016/j.optmat.2020.110198
51. T. Janusas, S. Urbaite, A. Palevicius, S. Nasiri, G. Janusas, Biologically compatible lead-free piezoelectric composite for acoustophoresis based particle manipulation techniques. Sensors **21**, 483 (2021). https://doi.org/10.3390/s21020483
52. H. Gheisari, E. Karamian, M. Abdellahi, A novel hydroxyapatite -Hardystonite nanocomposite ceramic. Ceram. Int. **41**, 5967–5975 (2015). https://doi.org/10.1016/j.ceramint.2015.01.033
53. M. Manoj, D. Mangalaraj, N. Ponpandian, C. Viswanathan, Core-shell hydroxyapatite/Mg nanostructures: Surfactant free facile synthesis, characterization and their in vitro cell viability studies against leukaemia cancer cells (K562). RSC Adv. **5**, 48705–48711 (2015). https://doi.org/10.1039/c5ra04663g
54. M.K.F. Lo, A. Dazzi, C.A. Marcott, E. Dillon, Q. Hu, K. Kjoller, C.B. Prater, S.W. King, Nanoscale chemical-mechanical characterization of nanoelectronic low- k dielectric/Cu interconnects. ECS J. Solid State Sci. Technol. **5**, P3018–P3024 (2016). https://doi.org/10.1149/2.0041604jss
55. I.R. Gibson, W. Bonfield, Novel synthesis and characterization of an AB-type carbonate-substituted hydroxyapatite. J. Biomed. Mater. Res. **59**, 697–708 (2002). https://doi.org/10.1002/jbm.10044
56. V. Dhand, K.Y. Rhee, S.J. Park, The facile and low temperature synthesis of nanophase hydroxyapatite crystals using wet chemistry. Mater. Sci. Eng. C **36**, 152–159 (2014). https://doi.org/10.1016/j.msec.2013.11.049

57. W.J. Shih, Y.F. Chen, M.C. Wang, M.H. Hon, Crystal growth and morphology of the nano-sized hydroxyapatite powders synthesized from CaHPO4·2H2O and CaCO3 by hydrolysis method. J. Cryst. Growth **270**, 211–218 (2004). https://doi.org/10.1016/j.jcrysgro.2004.06.023
58. T. Yan, W. Guan, J. Tian, P. Wang, W. Li, J. You, B. Huang, Improving the photocatalytic performance of silver phosphate by thermal annealing: Influence of acetate species. J. Alloy. Compd. **680**, 436–445 (2016). https://doi.org/10.1016/j.jallcom.2016.04.142
59. C.M. Murphy, M.G. Haugh, F.J. O'Brien, The effect of mean pore size on cell attachment, proliferation and migration in collagen-glycosaminoglycan scaffolds for bone tissue engineering. Biomaterials **31**, 461–466 (2010). https://doi.org/10.1016/j.biomaterials.2009.09.063
60. S.F. Hulbert, F.A. Young, R.S. Mathews, J.J. Klawitter, C.D. Talbert, F.H. Stelling, Potential of ceramic materials as permanently implantable skeletal prostheses. J. Biomed. Mater. Res. **4**, 433–456 (1970). https://doi.org/10.1002/jbm.820040309
61. J.J. Klawitter, J.G. Bagwell, A.M. Weinstein, B.W. Sauer, J.R. Pruitt, An evaluation of bone growth into porous high density polyethylene. J. Biomed. Mater. Res. **10**, 311–323 (1976). https://doi.org/10.1002/jbm.820100212
62. V. Karageorgiou, D. Kaplan, Porosity of 3D biomaterial scaffolds and osteogenesis. Biomaterials **26**, 5474–5491 (2005)
63. M. Manoj, R. Subbiah, D. Mangalaraj, N. Ponpandian, C. Viswanathan, K. Park, Influence of growth parameters on the formation of hydroxyapatite (HAp) nanostructures and their cell viability studies. Nanobiomedicine **2**, 2 (2015). https://doi.org/10.5772/60116
64. L.T. Bang, B.D. Long, B.D.R. Othman, Carbonate hydroxyapatite and silicon-substituted carbonate hydroxyapatite: synthesis, mechanical properties, and solubility evaluations. Sci. World J.**2014** (2014). https://doi.org/10.1155/2014/969876
65. K. Prasad, O. Bazaka, M. Chua, M. Rochford, L. Fedrick, J. Spoor, R. Symes, M. Tieppo, C. Collins, A. Cao et al., Metallic biomaterials: Current challenges and opportunities. Materials **10**, 884 (2017). https://doi.org/10.3390/ma10080884
66. A. Esmaeilkhanian, F. Sharifianjazi, A. Abouchenari, A. Rouhani, N. Parvin, M. Irani, Synthesis and characterization of natural nano-hydroxyapatite derived from turkey femur-bone waste. Appl. Biochem. Biotechnol. **189**, 919–932 (2019). https://doi.org/10.1007/s12010-019-03046-6
67. M.H. Fathi, A. Doost Mohammadi, Preparation and characterization of sol-gel bioactive glass coating for improvement of biocompatibility of human body implant. Mater. Sci. Eng. A **474**, 128–133 (2008). https://doi.org/10.1016/j.msea.2007.05.041
68. D.U. Tulyaganov, S. Agathopoulos, P. Valerio, A. Balamurugan, A. Saranti, M.A. Karakassides, J.M.F. Ferreira, Synthesis, bioactivity and preliminary biocompatibility studies of glasses in the system CaO-MgO-SiO2-Na2O-P2O 5-CaF2. J. Mater. Sci. Mater. Med. **22**, 217–227 (2011). https://doi.org/10.1007/s10856-010-4203-5
69. C. Guzmán Vázquez, C.P. Barba, N. Munguía, Stoichiometric hydroxyapatite obtained by precipitation and sol gel processes. Rev. Mex. Fis. **51**, 284–293 (2005)
70. S. Nasiri, M. Nasr-Esfahani, Bioactive organic-inorganic composite monolith derived from poly vinyl trimethoxy silane using sol-gel process. Plast. Polym. Technol. **2**, 63–67 (2013)
71. X.Y. Lü, Y. Fan, B. Gu, D. W. Cui, Preparation and characterization of natural hydroxyapatite from animal hard tissues. Key Eng. Mater. **342–343**, 213–216 (2007). https://doi.org/10.4028/www.scientific.net/kem.342-343.213
72. A. Ruksudjarit, K. Pengpat, G. Rujijanagul, T. Tunkasiri, Synthesis and characterization of nanocrystalline hydroxyapatite from natural bovine bone. Curr. Appl. Phys. **8**, 270–272 (2008). https://doi.org/10.1016/j.cap.2007.10.076
73. G. Lutzweiler, A.N. Halili, N.E. Vrana, The overview of porous, bioactive scaffolds as instructive biomaterials for tissue regeneration and their clinical translation. Pharmaceutics **12**, 1–29 (2020)
74. H. Wang, J. Pieper, F. Péters, C.A. van Blitterswijk, E.N. Lamme, Synthetic scaffold morphology controls human dermal connective tissue formation. J. Biomed. Mater. Res. Part A **74A**, 523–532 (2005). https://doi.org/10.1002/jbm.a.30232

75. Use of bioactive glass compositions to stimulate osteoblast production. U.S. Patent Application No. 10/332,731, 11.07.2001
76. M. Bohner, J. Lemaitre, Can bioactivity be tested in vitro with SBF solution? Biomaterials **30**, 2175–2179 (2009). https://doi.org/10.1016/j.biomaterials.2009.01.008
77. H.M. Kim, T. Himeno, M. Kawashita, T. Kokubo, T. Nakamura, The mechanism of biomineralization of bone-like apatite on synthetic hydroxyapatite: an in vitro assessment. J. R. Soc. Interface. **1**, 17–22 (2004). https://doi.org/10.1098/rsif.2004.0003
78. M. Ansari, S. Morteza Naghib, F. Moztarzadeh, A. Salati, Synthesis and characterization of hydroxyapatite-calcium hydroxide for dental composites. Ceramics **55**, 123–126 (2011)

Chapter 7
Influence of Calcination Temperature on the Photophysical and Mechanical Characteristics of Hydroxyapatite Doped with 5 mol% Copper Iodide

For the first time, a nanocomposite of 5 mol% crystalline CuI-doped hydroxyapatite (HA) was synthesized using a simple chemical technique and subjected to calcination at various temperatures, namely 300, 500, 700, and 900 °C. In this composition, HA served as the matrix, while CuI functioned as the reinforcement [1]. This study examined how different calcination temperatures influenced the XRD patterns, TEM, SEM, BET, FTIR, luminescence intensity, photoluminescence quantum yield (PLQY), fluorescence lifetime, charge mobility measurements, electronic speckle pattern interferometry, and mechanical properties. The particle size, as determined through the modified Scherrer equation, displayed an increase from 66.64 to 87.20 nm as the calcination temperature rose. At room temperature, the photoluminescence intensity of CuI (5 mol%)/HA was reduced, whereas the constant decay time for the calcined material increased at 700 °C. Additionally, the CIE coordinates for CuI (5 mol%)/HA calcined at 700 °C were determined to be 0.39333, 0.18493. The charge mobility values of the nanocomposites were extracted by space-charge limited current (SCLC) method and the range of effective mobility was from 3.645×10^{-4} to $6.697 \times 10^{-4}\ cm^2V^{-1}\ s^{-1}$. Mechanical properties were fully discussed via ASTM-E9 standard and CuI (5 mol%)/HA calcined at 900 °C was in better range than other compounds, the σ_{yc} and hardness values were reported 7.32 Mpa and 40.81 HV respectively. Speckle interferometry was used to demonstrate that there was no large imperfection in the surface of each sample. The components were very sensitive to the calcination temperatures. In other words, the combination of CuI (5 mol%)/HA showed that the replacement of Cu ions with Ca was helpful to improve the photophysical and mechanical properties.

M. Rabiei et al., *Hydroxyapatite-Based Nanocomposites*, Springer Series in Chemical Physics 127, https://doi.org/10.1007/978-3-032-12583-5_7

7.1 Introduction

The constituents of a composite are typically divided into matrix and reinforcement. Both of these elements are essential for forming a composite. The role of the reinforcement is to provide support to the matrix, and it relies on the mechanical and physical characteristics of the materials. Hydroxyapatite (HA) is a multifaceted material that shows promise for use in creating composites. There are several methods to prepare artificial HA such as solid state reaction, mechanochemistry [2], sonochemistry [3], hydrothermal [4], sol–gel [5] and precipitation [6]. Chemical formula of HA is ($Ca_{10}(PO_4)_6(OH)_2$) [7, 8] and recently HA has been considered by researchers as a potential adsorbent for heavy metals due to its large specific surface area, high thermal and chemical stability and high ion exchange capacity [9, 10]. There are two crystal structures of HA: (1) hexagonal structure and (2) monoclinic structure. In order to improve the properties of HA, the researchers replaced the Ca^{2+} ions with other metal ions. The structure of HA consists of a quasi-compact packing of phosphate groups, with two types of Ca^{2+} ions, such as tunnel and parallel positions. One of the main characteristics of the HA structure is the ability of a large number of substitutions, which does not change the basis of the crystallographic structure. In addition, research on copper iodide (CuI) is very attractive because it has several advantages on the properties such as a wide large band gap, photosensitivity, diamagnetism behavior, as a dye in solar cells, and superionic conductor [11, 12]. Several metals, e.g. Ag^{+}, Sr^{2+}, Mn^{2+} and Fe^{3+} with different weight percentages have been used for doping in the HA by different methods and the evaluations on the properties have been brought in Ref [13–16]. Moreover, the Cu^{2+}–doped HA was prepared by the sol–gel method in research of Renaudin et al. [17]. According to the experimental observations, it was undeniable that the Cu^{2+}–doped HA structure leads to bonding, and Cu^{2+} cations are placed along the crystallographic hydroxyl sites through the formation of linear O–Cu–O units. Subsequent studies involving copper have shown significant similarities and have also revealed specific behaviors. Moreover, a higher doping level can be achieved with a reduction of Cu^{2+} to Cu^{+}, but this requires a higher sintering temperature [18–20]. In this case, prepared composites, which consisted of CuI (5 mol%)/HA, were very sensitive to the calcination temperature (from 25 °C to 900 °C). Our primary objective was the characterization of HA doped with 5 mol% CuI, particularly to enhance its photophysical and mechanical properties. The nanocomposite was synthesized via a modified procedure and subjected to calcination at varying temperatures. This study highlights the advantage of a straightforward, controlled approach for preparing the CuI (5 mol%)/HA nanocomposite. The findings demonstrated that the formation of CuI/HA at diverse calcination temperatures, with precise control over particle nanosize, is suitable for applications in various optical and mechanical domains.

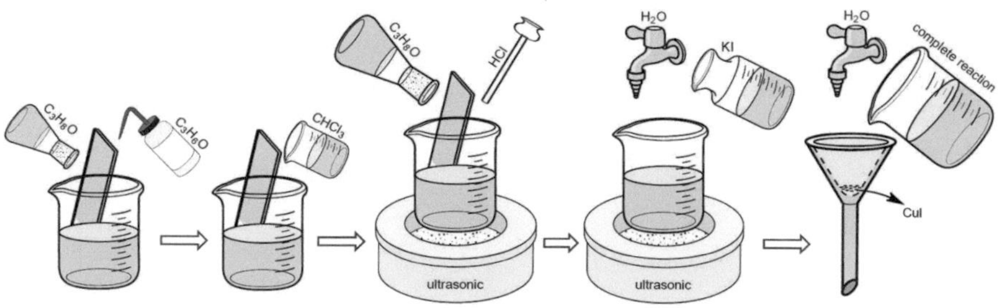

Scheme 7.1 Illustration of the synthesis method of CuI particles

7.2 Materials and Methods

7.2.1 Synthesis of CuI

In this subchapter, to achieve high uniformity and purity, copper (II) sulfate ($CuSO_4$) was coated on the clean quartz substrate by physical vapor deposition (PVD) method as a source of copper. As shown in Scheme 7.1, first, the copper substrate was immersed in acetone (C_3H_6O) and isopropyl alcohol (C_3H_8O), then the solution was etched and stirred with dilute aqueous chloroform ($CHCl_3$) (10 mol L^{-1}) for 20 min. Second, 5 ml of C_3H_8O and hydrochloric acid (HCl) were added to the solution under the vibrating ultrasound for 20 min; this procedure was repeated and filtered three times with deionized water, then the quartz substrate was removed. Third, 3 mmol solution of potassium iodide (KI) and 20 mL of deionized water were added and sonicated for 30 min, then the reaction was completed. Fourth, a gray precipitate was produced in the final step, and the product was washed, filtered, and dried at room temperature for four days. Furthermore, the summarized synthesis method of CuI particles is shown in Scheme 7.1. Additionally, CuI was prepared as a homogeneous reinforcement and to demonstrate the crystal phase of the particles, the XRD pattern is presented in Fig. 7.1.

7.2.2 Synthesis of CuI/HA Nanocrystalline

The reagents were calcium nitrate tetrahydrate $Ca(NO_3)_2.4H_2O$ as a calcium ion source and diammonium hydrogen phosphate ($(NH_4)_2HPO_4$) as a phosphorus precursor. In addition, concentrated nitric acid (HNO_3) and sodium hydroxide (NaOH) were added, which played a role in the dispersion of the particles, and copper iodide (CuI) was used as a dopant to prepare the CuI (5 mol%)/HA composite. In this procedure, nitric acid (HNO_3) and deionized water were added with the indicated amount of CuI until the particles were completely dissolved, and then heated until crystallization occurred. The unique mole fraction of $I_2 + 2Cu(NO_3)_2$ was obtained

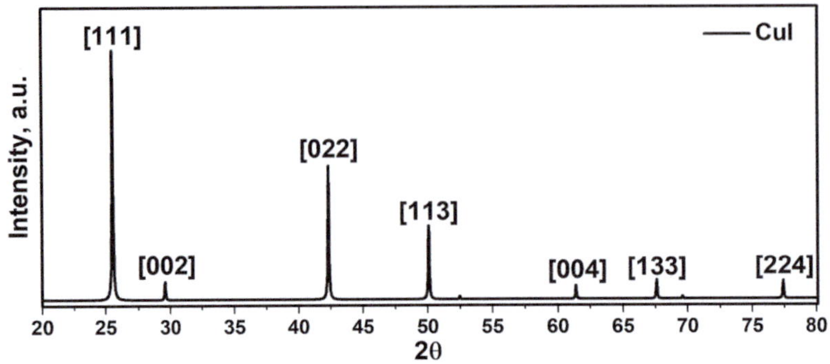

Fig. 7.1 XRD pattern of CuI

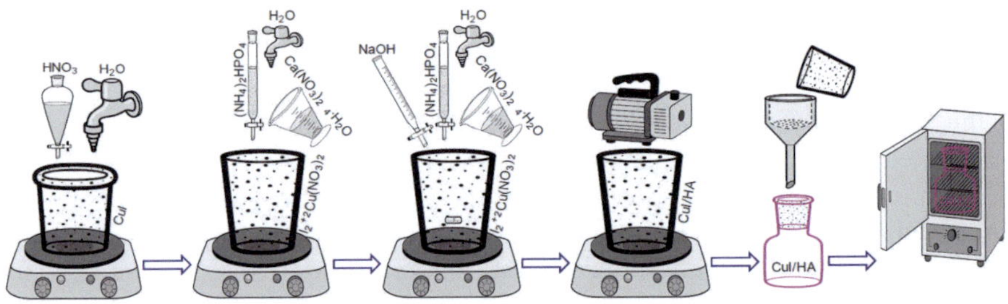

Scheme 7.2 Illustration of the synthesis method of CuI (5 mol%)/HA nanoparticles

and after dissolution of the crystals, the mole fraction of CuI was 5 mol%. In the next step, according to the molar ratio of (Ca + CuI)/P = 1.67, $(NH_4)_2HPO_4$ was added dropwise to the mixture of $Ca(NO_3)_2.4H_2O$ and $I_2 + 2Cu(NO_3)_2$ (with stirring at 250 rpm) and pH was achieved to 14 by adding 2 mol l^{-1} NaOH at room temperature. The product was dried at 100 °C and 0.76 bar simultaneously, and after one day the mixture was washed, filtered, and dried at 150 °C for 4 h. Finally, the compounds were calcined at 300 °C, 500 °C, 700 °C and 900 °C for 2 h. The summary of the synthesis method of CuI (5 mol%)/HA is shown in Scheme 7.2, and the synthesis routes of CuI and CuI (5 mol%)/HA are shown in Scheme 7.3.

7.2.3 *Characterization*

The phase sequence was validated using X-ray diffraction (XRD), conducted on a Philips XRD diffractometer. Cukα radiation was applied with settings of 40 kV and 30 mA, featuring a 0.05° (2Θ) step size and a 1°/min scan rate. X'Pert software (Version 4.9.0) aided in the qualitative analysis, documenting the width of diffraction peaks (rad, β) at various full width half maximum (FWHM) 2θ values,

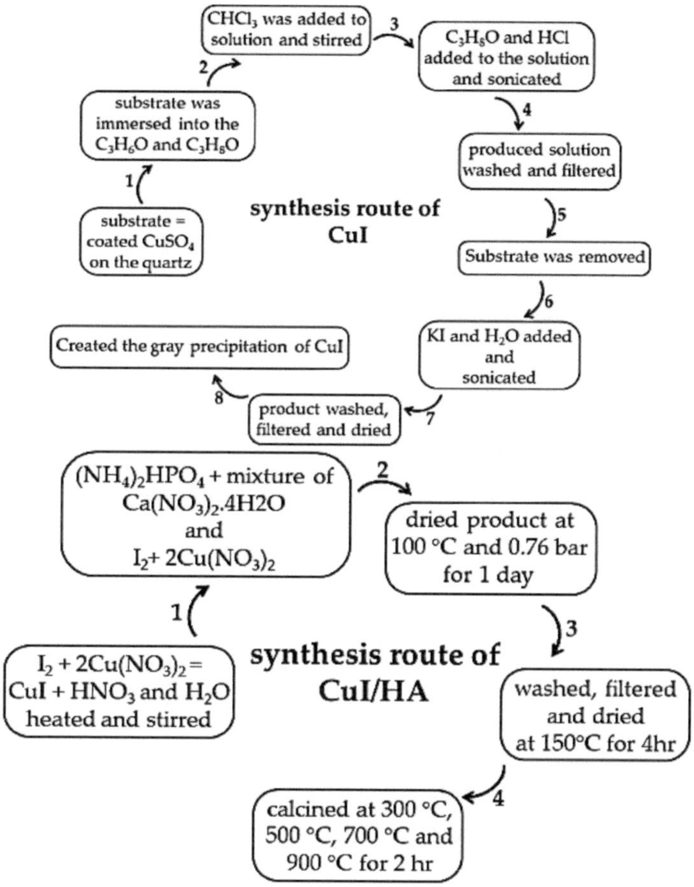

Scheme 7.3 The synthesis routes of CuI and CuI (5 mol%)/HA

based on peak positions. Additionally, transmission electron microscopy (TEM) was performed with a Tecnai G2 F20 X-TWIN, operating at an acceleration voltage between 50 and 80 kV. The specific surface area of the samples was assessed using nitrogen (N2) gas desorption isotherms measured by a Brunauer–Emmett–Teller (BET) apparatus with a Gemini V analyzer from Micrometrics GmbH. Furthermore, compounds were analyzed using Fourier transform infrared spectroscopy (FTIR) by incorporating them into potassium bromide (KBr) powders, using a Perkin-Elmer Spectrum BX FT-IR spectrometer. UV–visible spectroscopy was performed with an AvaSpec-ULS2048XL-EVO and AvaSoft 8 software. Photoluminescence spectra of the solid films of compounds were captured using a Hitachi F-4600 FL spectrophotometer over the range of 400–800 nm, with a Xenon discharge lamp providing a light source within 250 to 1300 nm. The photoluminescence decay spectrum was measured using a PicoQuant laser along with its instrument response (IR), at a wavelength of 375 nmnm. The response function of the instrument captures the initial value with a slow constant time, indicating an exceptionally rapid energy transfer rate. Infrared (IR) is typically noted as reflected or scattered excitation light, possessing a shorter wavelength compared to the fluorescence emission that is observed. To acquire the IR synchronized with the laser pulse and detection response, an aqueous erythrosine solution was employed in both time-resolved fluorescence and delayed fluorescence experiments. Furthermore, thermal gravimetric analysis (TG) and differential scanning calorimetry (DSC) were conducted using a TA Instruments Q2000 thermal system. These analyses were performed on compounds in a nitrogen atmosphere, with a heating rate of 10 °C/min. For component chemical element analysis, an energy dispersive X-Ray (EDX) spectrometer from Phillips/FEI Quanta 200 was used. Additionally, zeta potential measurements were conducted using an electroacoustic technique with an electro kinetic sonic amplitude (ESA) measurement apparatus acoustosizer from Colloidal Dynamics. The glove box (vacuum 1×10^{-6} torr) was used and devices for mobility measurements were fabricated through the physical vapor deposition (DS1-170,PVD) method. Furthermore, for calculating values of mobility, the space charge limited current (SCLC) technique was performed. Current density–voltage (J–V) experiments were performed on the devices with a source measure unit model of Keithley 238 at room temperaturetemperature. Additionally, for assessing mechanical properties, the SANTAM STM-50 and H-25KT devices were employed, with specimens being prepared following the ASTM-E9 standard [21]. Furthermore, hardness measurements were conducted using the LECO M-400 model, in accordance with the ASTM-E384 standard [22]. The high-tech Electronic Speckle Pattern Interferometry (ESPI) system PRISM (MM-SIC) alongside the PRISM VIZ (DAQ-Version 1.2.0) software was utilized for examining surface imperfections on specimens. Also, a scanning electron microscope (SEM) Phillips/FEI Quanta 200 was used to analyze the morphology of the compounds.

7.3 Results and Discussion

7.3.1 X-Ray Diffraction and TEM Analysis

The XRD patterns for CuI and CuI (5 mol%)/HA, after calcination at varied temperatures, are depicted in Figs. 7.1 and 7.2a. The pattern for CuI aligns with the standard specified in X'Pert software and references [23, 24]. As shown in Fig. 7.1, the reflection peaks are indicative of copper iodide's cubic crystal phase, with no impurity peaks detected in the diffraction pattern. The pattern for CuI (5 mol%)/HA matches well with the standard HA XRD pattern in [25], showing a pure phase of CuI/HA at varying calcination temperatures. In Fig. 7.2a, it is noted that the intensity of the diffraction patterns rises with the increase in calcination temperature, which corresponds to particle size growth at elevated temperatures [26]. The compounds' diffraction patterns display pure peaks without any new stable phases; the peaks at [111] and [022] (notably at 900 °C) are associated with CuI and are attributed to the lower content of 5 mol% CuI-doped HA. In addition, peaks associated with the conversion of HA to the α-tricalcium phosphate (α-TCP), β-tricalcium phosphate (β-TCP) and tetra-tricalcium phosphate (T-TCP) can be reported [27]. Although the intensity of the compounds that had a slight shift of the peaks to the higher 2Θ is due to the substitution and exchange of the larger size of Ca^{2+} ions with the smaller size of Cu^{2+} ions, the increase of intensity was mainly proportional to the increase in the degree of crystallinity of CuI/HA. In addition, lnβ versus ln(1/cosΘ) was represented by the equation of linear least squares method obtained from the linear regression of the data in the original plot, and this type was cited by Ref [25]. According to the modified Scherrer equation, $\left(\mathrm{Ln}\beta = \mathrm{Ln}\left(\frac{\mathbf{K}\lambda}{\mathbf{L}}\right) + \mathrm{Ln}\left(\frac{1}{Cos\theta}\right)\right)$, β (FWHM) and Θ (Bragg angle) were converted to radians (π/180) and Eq. 7.1 was used to determine the particle size [28–30]. In the Monshi-Scherrer method, when Lnβ is plotted versus Ln(1/cosθ) and the method of least squares is applied, the intercept gives Ln(Kλ/L), from which a single value for L can be obtained [25, 28]. Furthermore, plots of the modified Scherrer of calcined compounds at different temperature are depicted in Fig. 7.3.

$$\frac{\mathrm{K}\lambda\mathrm{Cu}_{\mathrm{k}\alpha 1}}{\mathrm{L}} = \mathbf{e}^{(\text{intercept})} \tag{7.1}$$

In Eq. 7.1, K is the shape factor (K = 0.89), $\lambda_{\mathrm{Cuk}\alpha 1} = 0.15405$ nm, L is the crystal size, and the intercept is related to the linear equation. The linear equations of CuI

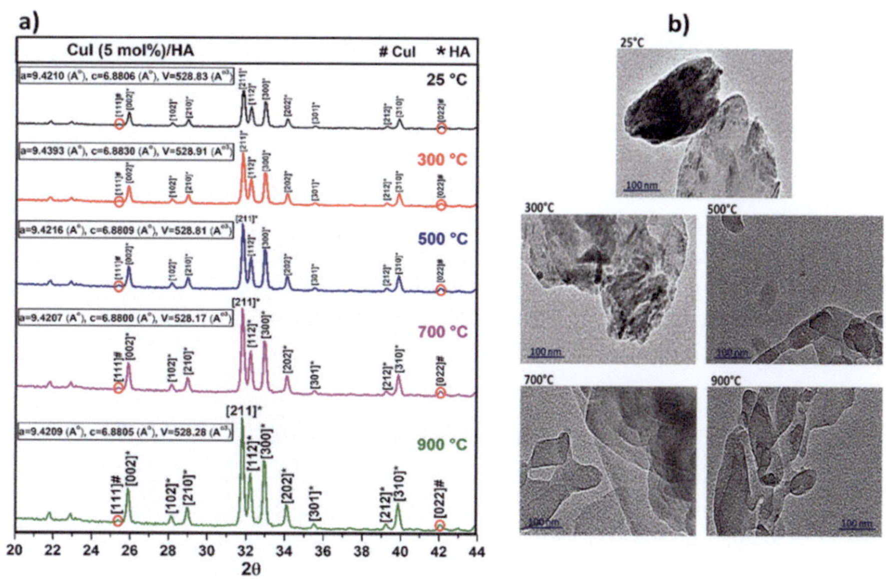

Fig. 7.2 **a** XRD pattern of CuI (5 mol%)/HA at 25 °C, calcined at 300, 500, 700 and 900 °C and **b** TEM images of compounds

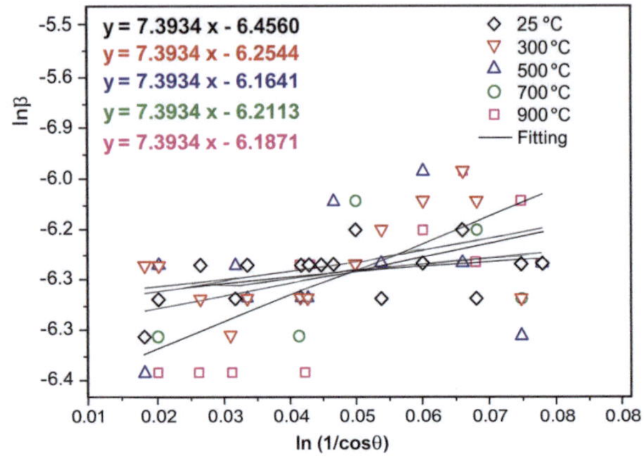

Fig. 7.3 Linear plots of the modified Scherrer equation of CuI (5 mol%)/HA at 25 °C, calcined at 300 °C, 500 °C, 700 °C and 900 °C

(5 mol%)/HA at 25°C, 300°C, 500 C, 700°C and 900 °C gave $y = 7.3934x - 6.4560$; $y = 7.3934x - 6.2544$; $y = 7.3934x - 6.1641$; $y = 7.3934x - 6.2113$ and $y = 7.3934x - 6.1871$ respectively. Nevertheless, the intercept is calculated as $e(-6.4560) = 0.00157$, $e(-6.2544) = 0.00192$, $e(-6.1641) = 0.00210$, $e(-6.2113) = 0.00201$ and $e(-6.1871) = 0.00206$ tandemly [29]. Therefore, the crystal size of CuI (5 mol%)/HA at 25 °C and calcined at 25, 300, 500, 700 and 900 °C, were calculated of 66.64, 68.27, 65.12, 71.28 and 87.20 nm respectively. Overall, according to the lower content, we did not observe the sharp peaks of CuI in Fig. 7.2a and also the reason is related to the behavior of molecules in different temperatures of calcination that the control of HA and TCP especially β-TCP (β-Tri Calcium Phosphate) phase was more convenient than CuI. On the other hand, the mineral composition of BCP and the actual position of CuI were dependent on doping CuI and the applied sintering temperature. Moreover, in this case, the substitution of CuI with BCP (Beta Calcium Phosphate) was not highlighted as an effect of sintering temperature on the HA/β-TCP ratio, similarly as this matter was mentioned in the researches of Renaudin et al.and Gomes et al. [31, 32]. Furthermore, TEM images of compounds are shown in Fig. 7.2b. According to the images, the size of particles basically corresponds to the crystallite size, and it is clear that particles have nano-scale and the size can be reported less than 100 nm.

7.3.2 Specific Surface Area by Gas Adsorption (BET Method)

The most commonly used method for estimating specific surface area is BET. Under normal atmospheric pressure and at the boiling temperature of liquid nitrogen, the amount of nitrogen adsorbed as a function of pressure gives the specific surface area of the powder. The observations are interpreted according to the model of the BET method. The theoretical particle size can be calculated from the specific adsorption surface area data by using and following Eq. 7.2.

$$D = \frac{6}{\rho.S} \tag{7.2}$$

In this formula, **ρ** is the density of the sample and S refers to the specific surface area of the sample determined by the BET method [33]. The specific surface area is ~ 178.6 m^2/gr and the particle size values for CuI (5 mol%)/HA at 25 °C, calcined at 300, 500, 700 and 900 °C were recorded ~53.89, 57.21, 63.75, 70.03 and 81.43 nm respectively and the values were in accordance with the size values extracted by XRD patterns.

7.3.3 FTIR Spectroscopy

The FTIR spectra of CuI (5 mol%)/HA, calcined at different temperatures, including the characteristic bands of the structures, are shown in Fig. 7.4. The broad peak centered around 3450 cm^{-1} is attributed to the bond between H and O and adsorbed water from the hydroxyl group [34, 35], incidentally, this area can also be associated with overlapping the strong bands. The observed spectra at wavelength from 580 to 603 cm^{-1} are due to the bonding between P and O associated with the crystalline phosphate phase like HA, whose existence HA was confirmed by the XRD pattern. A strong band at 1035 cm^{-1} and weak bands close to 1452 cm^{-1} and 1625 cm^{-1} were shown that the compounds are composed of phosphate and carbonate from HA, and it has been reported by Wang and Murugan et al. [36, 37].

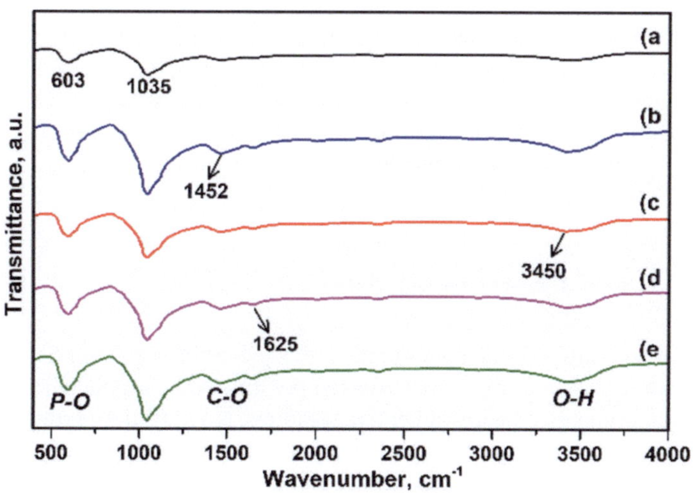

Fig. 7.4 FTIR spectra of **a** CuI (5 mol%)/HA at 25 °C, calcined at **b** 300 °C, **c** 500 °C, **d** 700 °C and **e** 900 °C

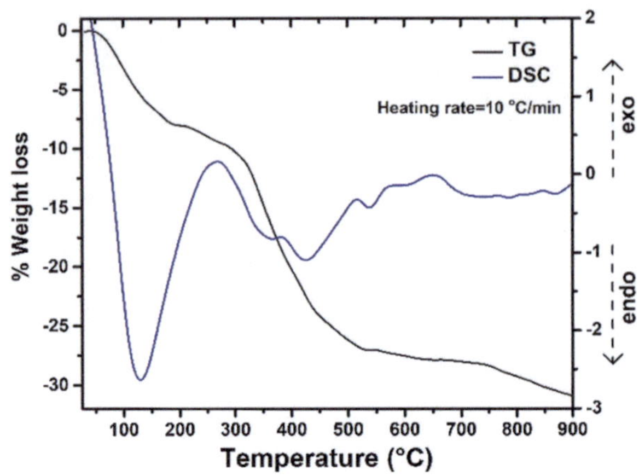

Fig. 7.5 TG and DSC curves of CuI (5 mol%)/HA at 25 °C

7.3.4 *Thermal Properties*

The TG and DSC curves for CuI (5 mol%)/HA at 25 °C are depicted in Fig. 7.5. The DSC curve reveals both significant and minor endothermic and exothermic events. Notably, pronounced endothermic peaks are recorded from room temperature up to 137 °C, indicative of water loss as verified by TG analysis. The TG curve indicates an overall weight reduction of approximately 27 wt%. This weight reduction occurs between roughly 200 °C and 550 °C, likely due to the decomposition of organic materials [38]. Additionally, an exothermic peak for CuI (5 mol%)/HA at 25 °C appears around 670 °C, not linked to any weight loss, and is attributed to a phase change in HA, as suggested by the XRD pattern of CuI (5 mol%)/HA that was calcined at 700 °C [39]. Further, after 700 °C, the TG curve presents mass loss alongside weak exothermic and endothermic bands, which is ascribed to the decarbonization process; this can persist up to 900 °C with varying intensities depending on the sample source, as well as the commencement of the dehydroxylation process [40].

7.3.5 *SEM and Edax-Energy Analysis*

Moreover, SEM images of the compounds are depicted in Fig. 7.6. It is evident that the particle size enlarges as the calcination temperature rises, and the visibility of CuI becomes more pronounced with higher temperatures. Notably, the aggregation of CuI (5 mol%)/HA is more visible at calcination temperatures of 700–900 °C, attributed to the nucleation and growth interactions between CuI and HA particles. Additionally, cross-sectional, clustered, and needle-like shapes of the particles are not observed. Elemental presence was confirmed by Edax analysis, with results shown in Table 7.1, revealing that element content is near the statistical detection threshold. The analysis indicates that the compound's inorganic phases primarily consist of elements Ca and

P, along with trace elements such as Cu, I, and O. From the Edax data, the (Ca + Cu + I)/P molar ratio of CuI (5 mol%)/HA at 25 °C was calculated to be 2.03, aligning with previously reported values in Ref [41, 42]. Essentially, the introduction of Cu and I ions leads to a depletion of Ca ions. In this case, an increased depletion of Ca ions is distinctly observed, which is linked to ionic substitution processes occurring through crystal defects among Cu, I, and Ca ions.

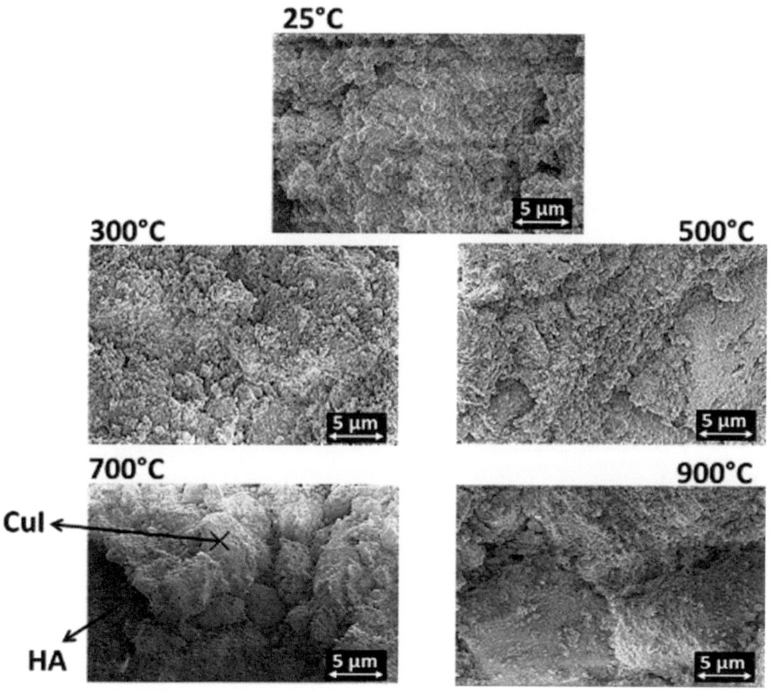

Fig. 7.6 SEM images of compounds

Table 7.1 EDX analysis of CuI (5 mol%)/HA at 25 °C

Element	25 °C	300 °C	500 °C	700 °C	900 °C
	% mol	% mol	% mol	% mol	% mol
O	54.61 ± 3.08	49.54 ± 4.18	56.02 ± 3.11	55.20 ± 2.28	54.27 ± 3.26
P	14.66 ± 1.17	15.29 ± 2.26	14.39 ± 1.87	15.44 ± 1.15	14.61 ± 1.27
Ca	24.77 ± 1.63	29.14 ± 1.98	23.11 ± 1.93	23.16 ± 1.06	24.77 ± 1.99
I	1.04 ± 0.18	1.01 ± 0.11	1.56 ± 0.12	1.21 ± 0.21	1.35 ± 0.14
Cu	4.07 ± 0.06	4.19 ± 0.08	4.01 ± 0.03	4.19 ± 0.09	4.23 ± 0.03
Cu ®	0.85 ± 0.09	0.83 ± 0.06	0.91 ± 0.09	0.80 ± 0.03	0.77 ± 0.01

7.3.6 Characterization of Water Suspensions of CuI (5 mol%)/HA Powder

The isoelectric point of CuI (5 mol%)/HA dispersed in H_2O at 25 °C was investigated by measuring the zeta potential and pH parameters; moreover, the compound was prepared by a classical precipitation technique via the neutralization route and previously reported in preto et al. [43]. The measurements of zeta potential versus conductivity as a function of pH are shown in Fig. 7.7. The conductivity of the suspension shows a sharp increase attributed to the dissolution event of CuI/HA. $\mathbf{H^+}$, $\mathbf{OH^-}$ and lattice ions have been shown to be potential ions for HA [44], the result of pH is the surface charge and adsorption or dissolution of calcium and phosphate ions. The zeta potential shifting towards negative values for CuI (5 mol%)/HA at 25 °C results from the substantial adsorption of hydrogen phosphate (HPO4) and iodine (I) ions. This particular anionic adsorption leads to a decrease in pH, moving it towards more acidic levels. Importantly, this pH shift was not influenced by particles formed through neutralization [45, 46]. Conversely, the zeta potential value signifies the extent of repulsion between nearby particles bearing the same charge, with higher zeta potential correlating to greater particle stability. A low potential means that attraction surpasses repulsion, causing particle aggregation [47].

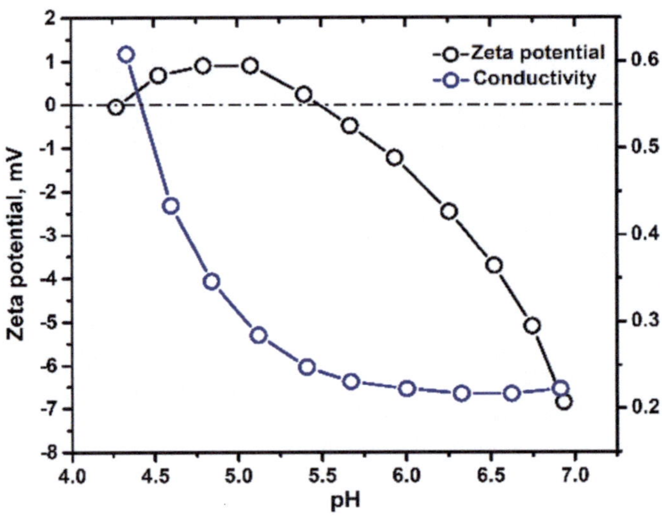

Fig. 7.7 Electroacoustic analysis of CuI (5 mol%)/HA at 25 °C

7.3.7 Absorption, Photoluminescence Intensity and Fluorescent Lifetime

Figure 7.8 illustrates the absorption spectrum for the compounds in their solid state, while Table 7.2 contains the corresponding maximum absorption wavelengths. These wavelengths range between 301 and 346 nm. Notably, in the longer wavelength region near 346 nm, narrow spectra are evident for CuI (5 mol%)/HA that underwent calcination at 700 °C, likely due to the excitation of electrons from valence subbands to the conduction band [48]. Interestingly, the reduction of Cu2 + to Cu + is identified in Fig. 7.8 through the red-shifted spectra of CuI (5 mol%)/HA, calcined at 900 °C, at 3.18 eV (Table 7.2). The values of $E_g^{optical}$ are close to each other and can be explained by the decrease in the ratio of surface area to volume as the crystals become larger, since the localized states most likely arise from surface states and this issue confirms that $E_g^{optical}$ is directly proportional to the size of the particles. Furthermore, it was interesting to note that the additional bands at around 344 and 346 nm were observed in Fig. 7.8 for CuI (5 mol%)/HA calcined at 300 and 700 °C, respectively. The exciton-phonon interaction of these compounds is the focus here, and based on XRD patterns, it is not linked to dynamic disorder, as both CuI (5 mol%)/HA samples calcined at 300 °C and 700 °C showed non-amorphous structures [49]. Table 7.2 details PLQY values for the coated compounds on solid films, where CuI (5 mol%)/HA calcined at 700 °C demonstrated higher PLQY compared to other compounds. Figures 7.9 and 7.10 display both the non-normalized and normalized photoluminescence spectra of the compounds at various temperatures in the solid state. The PL spectrum indicates that the emission intensity of CuI (5 mol%)/HA calcined at 700 °C surpassed that of the other compounds, attributed to its enhanced crystallinity. Higher crystallinity of CuI within the HA matrix results in superior crystal growth. At 25 °C, CuI (5 mol%)/HA displayed lower PL intensity, likely due to hole (h +) trapping at CuI surface sites, crucial to electron (e-) recombination [50, 51]. The emission at around 416–420 nm can be attributed to free exciton recombination, suggesting the perfection of the γ-CuI crystal [52, 53]. Moreover, the emission from 716 to 719 nm corresponds to the CuI + HA complex and the deficit of iodine in the γ-CuI crystal, which is consistent with the Ref [54]. The region from 716 to 719 nm is close to the band gap emission and also the emission in this region may come from bound excitons of CuI + HA complex and donor–acceptor pairs. Moreover, according to Fig. 7.2a, it was clear that the lattice parameter (a), lattice height (c) and unit cell volume (V) decreased for CuI (5 mol%)/HA calcined at 700 °C (a = 9.4207 Å, c = 6.8800 Å and V = 528.17 $\mathring{A}^3$). Accordingly, the Cu ions in the CuI (5 mol%)/HA calcined at 700 °C were eager to substitute and overlap with the Ca ions and moving to the edge of this unit cell was more favorable, while the calculated occupancy values for the two Ca crystal positions clearly show that the Cu^{2+} ions have a greater affinity to be accommodated at the Ca site, as these reasons were also confirmed by Nikolić et al. and Nedelec et al., respectively [17, 55]. Moreover, the high electronegativity values of $I^- > Cu^{2+} > Ca^{2+}$ ions were caused by the formation of covalent bonds with hydroxyl groups. In the structure of HA, the hydroxyl groups are positioned in the channels surrounded by Ca cations; therefore, a larger amount of Cu^{2+} ions is detected at this site, regardless of their total content.

The photoluminescence (PL) spectrum for the compounds displayed two prominent bands at approximately 416 nm and 719 nm. The peak near 416 nm is linked to the recombination band gap of CuI. In the region of 416 nm, the PL spectra suggest that energy absorption is concentrated in the donor–acceptor pairs [56–58]. This localization is related to the luminescence center, which is crucial for the photon excitation in CuI [59, 60]. Previous studies have reported the maximum PL wavelength for CuI to be 420 nm [61, 62], which aligns with the PL spectrum of CuI/

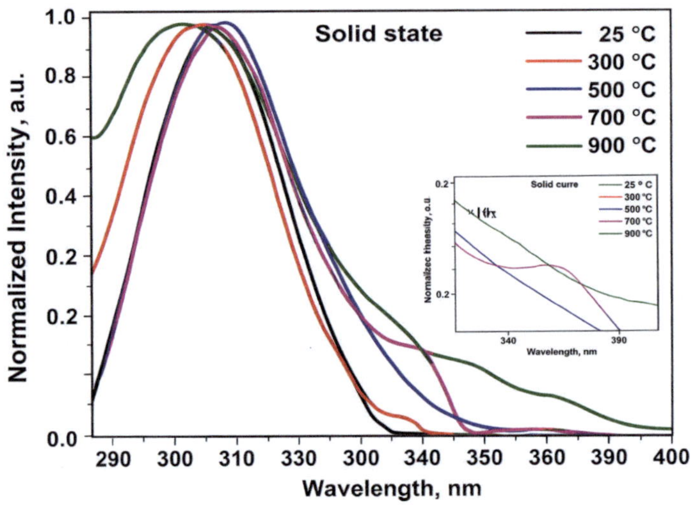

Fig. 7.8 Absorption spectra of compounds in solid film media

Table 7.2 Photophysical characteristics of compounds at different calcination temperatures

Compound (°C)	[a]$\lambda_{max\text{-}Abs}$ (nm)	[b]$\lambda_{max\text{-}PL}$ (nm)	[c]PLQY (%)	[d]$E_g^{optical}$ (eV)
25	307	420, 717	4.29	3.68
300	302, 344	419, 716	6.39	3.60
500	307	416, 718	6.51	3.48
700	308, 346	416, 719	19.08	3.51
900	301	417, 717	11.31	3.18

[a]Maximum wavelength of absorption
[b]Maximum wavelength of photoluminescence
[c]PLQY values of solid films
[d]Optical band gap Equation: 1240/λ ultraviolet spectra = fitting of long wavelength or end of absorption spectra

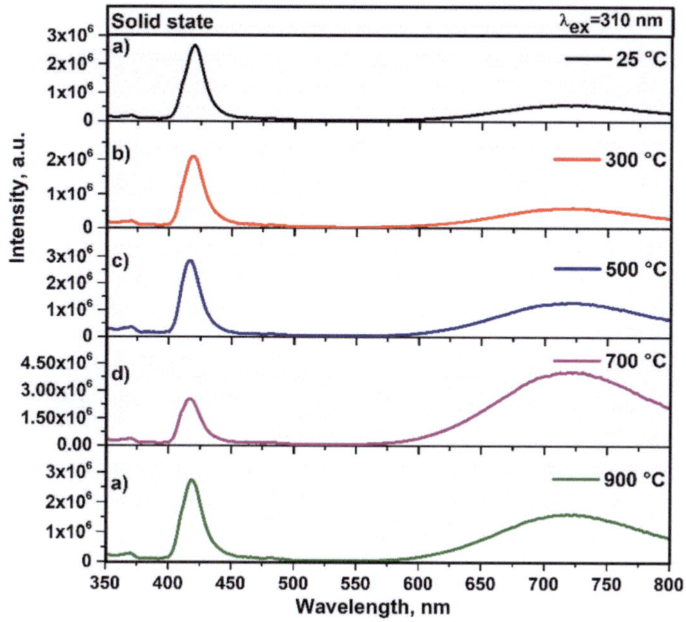

Fig. 7.9 Non-normalized photoluminescence spectra of **a** CuI (5 mol%)/HA at 25 °C, calcined at **b** 300 °C, **c** 500 °C, **d** 700 °C and **e** 900 °C in solid film media

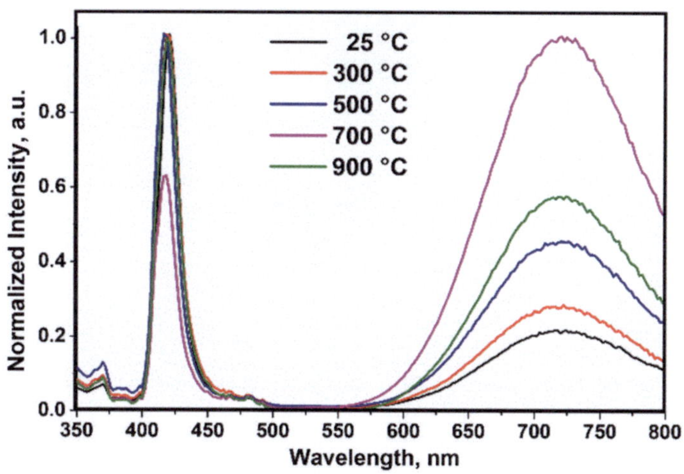

Fig. 7.10 Normalized photoluminescence spectra of **a** CuI (5 mol%)/HA at 25 °C, calcined at **b** 300 °C, **c** 500 °C, **d** 700 °C and **e** 900 °C in solid film media (ex = 310 nm)

HA at around 416 nm, reaffirming the recombination band gap of CuI. The calcination temperature's influence on the optical properties of the composites has been attributed to the diffusion of γ-CuI into the HA structure [63].

The Commission Internationale de l'Éclairage (CIE) chromaticity coordinates (x,y) for the compounds were demonstrated through the emission spectrum's peak wavelength in solid films (Fig. 7.11). Table 7.3 lists the specific CIE coordinate values. According to these coordinates, the color of the compounds transitioned to 171,472,632 Kelvin in the CIE 1931 system, particularly for CuI (5 mol%)/HA calcined at 700 °C. Additionally, a reversible color shift was noted as the calcination temperature increased to 900 °C, causing the color to change and then revert. This indicates that these compounds are highly sensitive to calcination temperature [64].

Figure 7.12 shows the spectra of count versus time, and the fitting parameters of lifetime as a function of calcination temperature of CuI (5 mol%)/HA are given in Table 7.3. The fluorescence and delayed fluorescence lifetimes of the compounds in the solid film media were plotted in a double exponential manner. Furthermore, for the compounds, the relaxation route was shown to proceed with a constant time

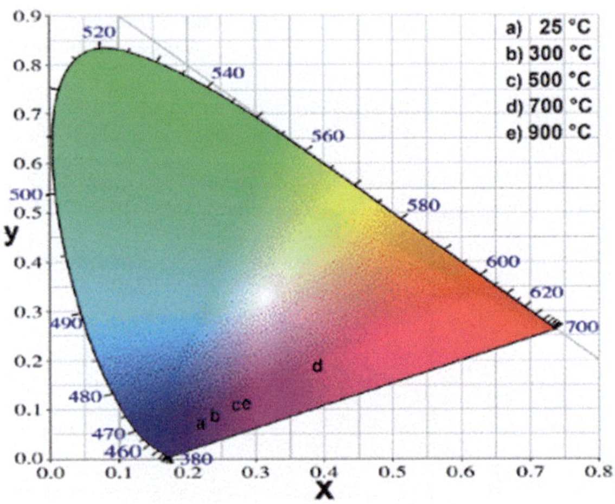

Fig. 7.11 CIE coordinates of **a** CuI (5 mol%)/HA at 25 °C, calcined at **b** 300 °C, **c** 500 °C, **d** 700 °C and **e** 900 °C

Table 7.3 Time life fitting data and CIE coordinate values of compounds in the different calcinations temperature

Compound (° C)	τ^{PF} (ns)	$\chi^2_{(PF)}$	τ_1^{DF} (μs)	τ_2^{DF} (μs)	Quantity of τ_1^{DF}(%)	Quantity of τ_2^{DF} (%)	$\chi^2_{(DF)}$	CIE coordinates (X,Y)
25	1.19	1.130	4.09	19.81	49.09	50.91	1.022	0.22421,0.07101
300	6.31	1.028	9.22	27.11	45.51	54.49	1.181	0.23500,0.08453
500	8.89	1.001	12.15	34.17	48.70	51.30	1.020	0.27733,0.10712
700	12.17	1.019	18.34	69.91	44.11	55.89	1.038	0.39333,0.18493
900	10.07	1.008	15.81	51.47	47.15	52.85	1.300	0.28167,0.10923

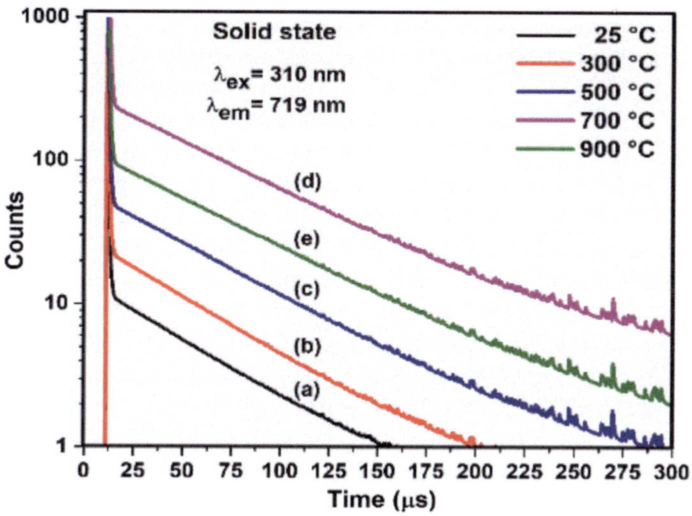

Fig. 7.12 Lifetime as a functional fitting curves of neat solid film of **a** CuI (5 mol%)/HA at 25 °C, calcined at **b** 300 °C, **c** 500 °C, **d** 700 °C and **e** 900 °C

in the range of microseconds, distinguishing the generation as prompt and delayed fluorescence. Equation 7.3 was used for the mathematical formula to fit the lifetime spectra.

$$I = I_0 + Ae^{-t/\tau} \tag{7.3}$$

where I is the representation of the fluorescence intensity at time (t), I_0 is the parameter of the intensity, A is the lifetime constant, t is the decay time, and τ is the prompt or delayed fluorescence lifetime [65]. Figure 7.12 illustrates that the PL decay increases at a calcination temperature of 700 °C and subsequently decreases at 900 °C. This phenomenon is associated with alterations in the molecular structure's geometry, which in turn affects the crystal field's influence on the matrix's energy levels, thereby enhancing the efficiency of delayed fluorescence. The interactions between neighboring atoms at the luminescence centers contribute to variations in transition processes, impacting the lifetime of excited state electrons through the atoms' coordination numbers. This concept is also explored in reference [66]. It is acknowledged that the emission at 420 nm is characterized by a short decay time, typically in the pico or nanosecond range [67, 68]. Furthermore, it has been demonstrated that the decay time for the emission band at 716 nm is significantly longer than that of the 420 nm emission.

The distinction in energy between the bottom edge of the conduction band and the top edge of the valence band accounts for the low and/or high energy emission bands. This energy is needed to liberate an electron residing in the outermost electron shell. Commonly, the forbidden band is referred to as the band gap or energy gap

[69, 70]. In this research, the source of the energy emission band is linked to the morphology of the compounds [71, 72] and singlet–triplet energies [73]. For CuI (5 mol%)/HA calcined at 700 °C, there is superior overlap between CuI and HA compared to other compounds, with CuI being completely and evenly distributed on HA (Fig. 7.6). Additionally, the crystal size of CuI (5 mol%)/HA at 700 °C is larger than at 25 °C, resulting in enhanced optical and emission properties. Furthermore, according to the values of PLQYs, represent constant time of PL decay and charge mobility, the strong charge transfer from the local excited triplet (^{3}LE) to the charge transfer singlet (^{1}CT) in CuI (5 mol%)/ HA calcined at 700 °C is very impressive. This leads to a small exchange interaction between the involved electrons and also to the required small splitting between the singlet (S_1) and triplet (T_1) states, therefore origin of the high-energy emission (^{3}LE $\rightsquigarrow$ ^{1}CT $\rightsquigarrow$ the ground state) is done, but in CuI (5 mol%)/ HA at 25 °C low-energy emission (^{3}LE $\rightsquigarrow$ the ground state) is done, since the triplet emits more weakly than the singlet [74, 75]. Chemical kinetics, lifetime and electronic states are directly related [76], nevertheless, according to the high values of τ^{PF}, $\tau_1{}^{DF}$ and $\tau_2{}^{DF}$ of CuI (5 mol%)/HA calcined at 700 °C the absorption of radiative energy exceeds the energy of the excited electronic state, HA molecule is left in an excited vibrational level of the excited electronic state, so the collision with the neighbouring CuI molecules rapidly relaxes it to the lowest vibrational level of that excited state, the excess vibrational energy is converted into heat and thermal equilibrium is obtained [77].

7.3.8 Charge Transporting Properties

Investigation of charge transport properties is one of the most important requirements for optical properties. In this case, we considered the space charge limited current (SCLC) method. After depositing layers according to the structure of ITO/ compounds (170 nm)/MoO_3 (1 nm)/Ag (250 nm), (Fig. 7.13a) devices were fabricated and according to the structure, electrons are blocked by a high LUMO (MoO_3) and only holes can pass through this structure. Moreover, after the lapse of different times such as 0 s, 86,400 s (1 day) and 172,800 s (2 days), the current density (CE) and effective mobility (μ_{eff}) were shown against the voltage characteristics of the devices in Fig. 7.13b. CuI (5 mol%)/HA calcined at 700 °C gives the highest current density compared to other devices. Moreover, an increase in applied voltage at t = 0 s leads to a rise in current density, a phenomenon linked to enhanced charge carrier mobility [78]. When CuI (5 mol%)/HA is calcined at 700 °C, it exhibits superior hole mobility compared to other compounds. This can be attributed to the compound's structural properties at this particular temperature, where the larger crystal size enhances the hole current facilitated by CuI [79]. Additionally, the Mott-Gurney law (Eq. 7.4) applies to the SCLC method.

$$J = \frac{9}{8}\varepsilon_0\varepsilon_r\mu_{eff}\frac{V^2}{d^3} \quad (7.4)$$

In this equation, J is the current density, ε_o is the vacuum permittivity, ε_r is the relative dielectric constant, d is the thickness of the sandwich devices, and V is the applied voltage, and as expressed in the formula, voltage and current density are proportional to effective hole mobility. The permittivity of the free space and relative dielectric constant are 8.85×10^{-14} and 3, respectively [80]. The values of effective hole mobility at t = 0 s, voltage = 10 V are reported of 3.645×10^{-4}, 4.029×10^{-4}, 5.632×10^{-4}, 6.697×10^{-4} and 6.532×10^{-4} cm^2/Vs for CuI (5 mol%)/HA at 25 °C, calcined at 300 °C, 500 °C, 700 °C and 900 °C, respectively (Table 7.4).

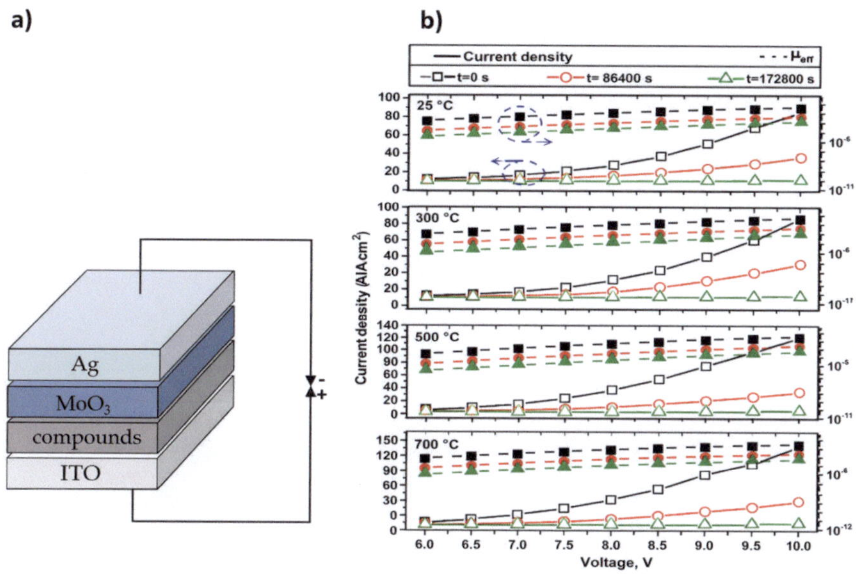

Fig. 7.13 **a** Fabricated devices and **b** Current density-effective mobility versus the voltage characteristics of

Table 7.4 Characteristics of hole and electron mobility of compounds extracted by SCLC method

Compound (° C)	μ^h_{eff} at t = 0 s, V = 10 V extracted by SCLC ($cm^2\ V^{-1}\ s^{-1}$)
25	3.645×10^{-4}
300	4.029×10^{-4}
500	5.632×10^{-4}
700	6.697×10^{-4}
900	6.532×10^{-4}

7.3.9 Mechanical Properties

The mechanical properties were examined using a compression test conducted at room temperature. The relative speed of the jaw was set to 0.1 mm/min, while the load cell registered 50 KN simultaneously. Figure 7.14 presents the stress–strain curve for both the composites and the powder compression system. Table 7.5 displays the mechanical test results along with the elasticity coefficient (the gradient of the elastic part). The elasticity coefficient is determined through Eq. 7.5.

$$E = \frac{\Delta\sigma}{\Delta\varepsilon} \tag{7.5}$$

Based on Eq. 7.5, $\Delta\sigma$ and $\Delta\varepsilon$ represent the differences in values at two specific points within the elastic region. Additionally, for the hardness measurements, a force of 300 g-force was applied, and the hardness was assessed at five different locations, with results detailed in Table 7.5. The microhardness averaged between 39.10 and 41.02 HV. Compression test results revealed that calcination at 900 °C enhanced the mechanical properties. Conversely, lower calcination temperatures resulted in

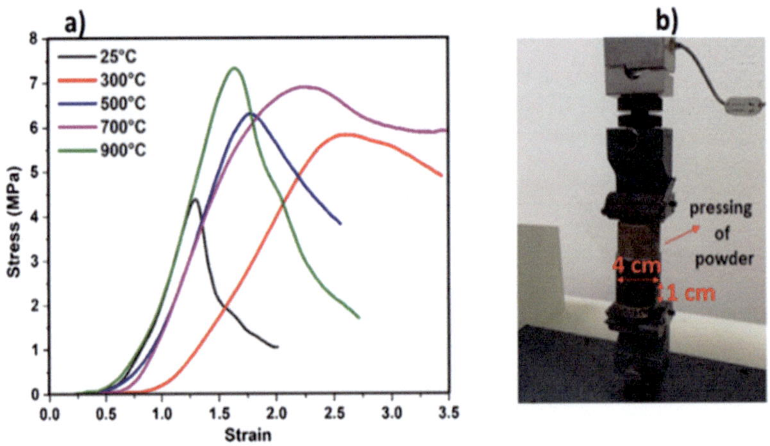

Fig. 7.14 **a** Compressive stress–strain curves of samples, and **b** Pressing of powder

Table 7.5 Values of σ_{yc} and ε extracted by compressive stress–strain curve of compounds

Compound (° C)	Hardness (HV)	E (GPa)	σ_{yc} (MPa)	ε in σ_{yc} (%)
25	39.54	4.68	4.35	1.29
300	39.21	**3.89**	**5.83**	2.59
500	41.02	3.11	6.29	1.76
700	39.10	4.09	6.89	2.22
900	40.81	3.95	7.32	1.63

reduced performance in compression and plasticity tests, attributed to the change in crystal size. As the crystal size increased, the intermixed phase of Cu and I expanded. Furthermore, the elasticity coefficient for all compounds was computed and ranged from 3.11 to 4.68 GPa, differing significantly from natural hydroxyapatite [81].

7.3.10 Digital Speckle Interferometry

Digital speckle interferometry is a prominent technique for identifying flaws in nondestructive analysis. This technique involves applying force to a sample, causing it to develop strain and stress patterns (interferometric lines) that can be interpreted. Typically, strain measurements are quite small, often on the order of micrometers. As a result, the integrity of the samples (through strain analysis) following mechanical processing was examined using the Speckle Pattern Interferometry (ESPI) system PRISM (refer to Fig. 7.15). During this evaluation, a green light laser (532 nm) with a power of 20 mW was employed. In Fig. 7.15, the object beam (3) illuminates the sample (5), and the scattered light collected by the camera (2) interferes with the reference beam, with the speckle pattern being captured by the CCD sensor and transmitted to the computer for further analysis. The speckle patterns are processed and displayed using the PRISM VIZ software.

Ensuring top-notch sample quality is crucial and depends on the volumetric distribution of the material, as well as the presence of internal fractures or air bubbles. The resulting images depicting the speckle pattern are presented in Fig. 7.16.

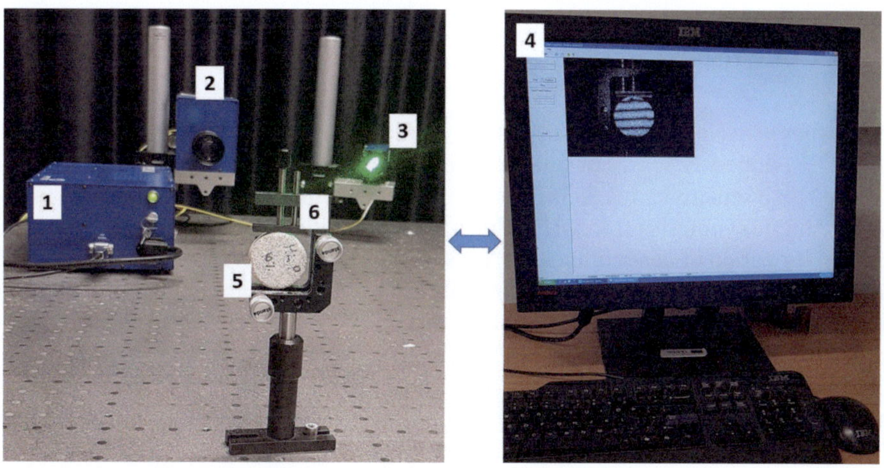

Fig. 7.15 Schematic configuration of speckle interferometry; electronic speckle pattern interferometry system PRISM consisted of; 1—control block; 2—video head; 3—illumination head, 4—monitor, 5—specimen, 6—holder

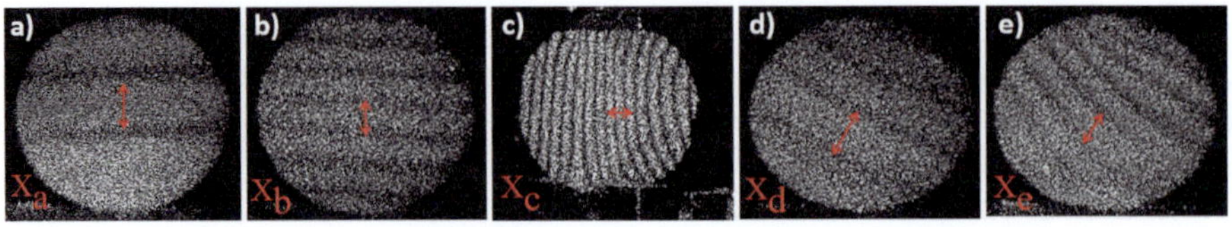

Fig. 7.16 Resulted speckle pattern interferometry images of specimens **a** CuI (5 mol%)/HA at 25 °C, calcined at **b** 300 °C, **c** 500 °C, **d** 700 °C and **e** 900 °C

Based on Fig. 7.16, samples exhibit uniform deformation under central loading with no visible defects, whereas parallel oblique lines suggest non-central loading. Interferometric lines with irregular shapes can detect defects, altering the sample's response to central loading. Local defects, such as air bubbles or cracks, decrease the elastic modulus, leading to concentric rings in these areas, indicating increased deformation. As interferometric lines become closer and denser, it signifies enlarged imperfections [82, 83]. From this analysis and Fig. 7.16, we assessed the spacing between the interferometric lines of the samples, and consequently $X_a > X_d > Xe \geq X_b > X_c$; therefore, the compound consisted of CuI (5 mol%)/HA at 25 °C and calcined at 700 °C have fewer imperfections than other compounds and it is clear that the compound at calcination temperature of 700 °C has similar distributions of stress and strain as CuI (5 mol%)/ HA at 25 °C. Nonetheless, this discussion relates directly to the surface of the samples and the speckle interferometry is entirely focused on the surface. Furthermore, according to the speckle images, there are no significant imperfections such as holes or cracks on the surface of the samples.

7.4 Conclusions

A comprehensive account was given of the straightforward synthesis of CuI (5 mol%)-doped HA, which retains its original crystal structure. The composites showed high sensitivity to calcination temperatures. This chapter introduces an innovative simple synthesis protocol for CuI, a novel technique for CuI-doped HA, and the initial fabrication of nanocomposites. By selecting an appropriate matrix like HA and a reinforcing agent such as CuI, it is possible to create a new compound that satisfies specific requirements of various applications, particularly those involving luminescence and photophysical properties. Additionally, the study thoroughly explored how calcination temperature impacts crystallite size, bonding, as well as photophysical and mechanical characteristics. The successful synthesis of CuI (5 mol%)-doped HA was confirmed by XRD pattern and Edax analysis. It was observed that in this research, the crystal size of the compounds increased with higher calcination temperatures, with crystallite size values ranging from 66.64 to 87.20 nm. Evidence for particle size was determined using surface area measurement through the BET test. The compound colors varied with calcination temperature, with the CIE coordinate values for CuI (5 mol%) at 700 °C recorded as 0.39333 and 0.18493. The

PL intensity and PLQY for CuI (5 mol%)/HA calcined at 700 °C exceeded those of other compounds, attributed to the recombination of electrons associated with CuI and ionization holes trapped within the Cu + HA complex, along with additional iodine introduced by CuI/HA during crystallization. The reported optical energy gaps of the compounds spanned from 3.18 to 3.68 eV. PL and PL decay measurements at 77 K using a cryostat system showed indistinct spectra. Furthermore, due to the absence of special components like Ir or Pt and the low 5 mol% CuI content, no phosphorescent effects were observed. The delayed fluorescence lifetime for CuI (5 mol%) calcined at 700 °C reached 69.91 μs. Compound charge mobility was analyzed using the SCLC method. The values of effective hole mobility were found to be in the range from 3.645×10^{-4} to 6.697×10^{-4} cm^2V^{-1} s^{-1}. The stress–strain curves of the samples were depicted and the maximum value of σ_{yc} was 7.32 MPa. Moreover, impressive imperfections were not observed on the surface of the samples according to the extracted images by speckle interferometry analysis. Finally, most of the properties increased with increasing calcination temperature and the optimum point of calcination temperature reached 700 °C; further, the CuI-doped HA was stabilized with almost no impurity phases present in the samples.

References

1. S. Nasiri, M. Hosseinnezhad, M. Rabiei, A. Palevicius, G. Janusas, The effect of calcination temperature on the photophysical and mechanical properties of copper iodide (5 mol%)–doped hydroxyapatite. Optical materials. Amsterdam. ISSN 0925-3467. eISSN 1873-1252. **121**(111559), 1–12. 10.1016/j https://www.sciencedirect.com/science/article/pii/S092534672100759X
2. C.W. Chen, C.S. Oakes, K. Byrappa, R.E. Riman, K. Brown, K.S. TenHuisen, V.F. Janas, Synthesis, characterization, and dispersion properties of hydroxyapatite prepared by mechanochemical-hydrothermal methods. J. Mater. Chem. **14**, 2425–2432 (2004). https://doi.org/10.1039/b315095j
3. G.E.J. Poinern, R.K. Brundavanam, X. Thi Le, P.K. Nicholls, M.A. Cake, D. Fawcett, The synthesis, characterisation and in vivo study of a bioceramic for potential tissue regeneration applications, Sci. Rep. **4** (2014). https://doi.org/10.1038/srep06235
4. C. Zhang, J. Yang, Z. Quan, P. Yang, C. Li, Z. Hou, J. Lin, Hydroxyapatite nano-and microcrystals with multiform morphologies: Controllable synthesis and luminescence properties. Cryst. Growth Des. **9**, 2725–2733 (2009). https://doi.org/10.1021/cg801353n
5. H. Zhou, J. Lee, Nanoscale hydroxyapatite particles for bone tissue engineering. Acta Biomater. **7**, 2769–2781 (2011). https://doi.org/10.1016/j.actbio.2011.03.019
6. H. Ito, Y. Oaki, H. Imai, Selective synthesis of various nanoscale morphologies of hydroxyapatite via an intermediate phase. Cryst. Growth Des. **8**, 1055–1059 (2008). https://doi.org/10.1021/cg070443f
7. R. Marzieh, N. Sohrab, P. Arvydas, J. Giedrius, Preparation and investigation of bioactive organic-inorganic nano-composite derived from PVB-co-VA-co-VAc/HA, in 15th *International Conference Mechatronic Systems and Materials MSM 2020, Institute of Electrical and Electronics Engineers Inc* (2020). https://doi.org/10.1109/MSM49833.2020.9201632
8. M. Rabiei, A. Palevicius, A. Dashti, S. Nasiri, A. Monshi, A. Doustmohammadi, A. Vilkauskas, G. Janusas, X-ray diffraction analysis and Williamson-hall method in USDM model for estimating more accurate values of stress-strain of unit cell and super cells (2 × 2 × 2) of

Hydroxyapatite, confirmed by ultrasonic pulse-echo test. Materials (Basel). **14**, 2949 (2021). https://doi.org/10.3390/ma14112949
9. S.A. Jahan, M.Y.A. Mollah, S. Ahmed, M.A.B.H. Susan, Copper-doped hydroxyapatite for removal of arsenic(V) from aqueous system. J. Sci. Res. **9**, 383–402 (2017). https://doi.org/10.3329/jsr.v9i4.32606
10. M. Rabiei, A. Palevicius, R. Ebrahimi-Kahrizsangi, S. Nasiri, A. Vilkauskas, G. Janusas, New approach for preparing in vitro bioactive scaffold consisted of Ag-doped hydroxyapatite + Polyvinyltrimethoxysilane. Polymers (Basel). **13**, 1695 (2021). https://doi.org/10.3390/polym13111695
11. M. Ferhat, A. Zaoui, M. Certier, J.P. Dufour, B. Khelifa, Electronic structure of the copper halides CuCl, CuBr and CuI. Mater. Sci. Eng. B **39**, 95–100 (1996). https://doi.org/10.1016/0921-5107(95)01518-3
12. V.N. Ermakov, S. Kruchinin, Electronic nanosensors based on nanotransistor with bistability behaviour (n.d.). https://doi.org/10.1007/978-1-4020-9146-9_26
13. Y. Huang, Z. Xu, X. Zhang, X. Chang, X. Zhang, Y.C. Li, T. Ye, R. Han, S. Han, Y. Gao, X. Du, H. Yang, Nanotube-formed Ti substrates coated with silicate/silver co-doped hydroxyapatite as prospective materials for bone implants. J. Alloys Compd. **697**, 182–199 (2017). https://doi.org/10.1016/j.jallcom.2016.12.139
14. A. Dubnika, D. Loca, V. Rudovica, M.B. Parekh, L. Berzina-Cimdina, Functionalized silver doped hydroxyapatite scaffolds for controlled simultaneous silver ion and drug delivery. Ceram. Int. **43**, 3698–3705 (2017). https://doi.org/10.1016/j.ceramint.2016.11.214
15. C.S. Ciobanu, S.L. Iconaru, F. Massuyeau, L.V. Constantin, A. Costescu, D. Predoi, Synthesis, structure, and luminescent properties of europium-doped hydroxyapatite nanocrystalline powders. J. Nanomater. (2012). https://doi.org/10.1155/2012/942801
16. M. Mirzaee, M. Vaezi, Y. Palizdar, Synthesis and characterization of silver doped hydroxyapatite nanocomposite coatings and evaluation of their antibacterial and corrosion resistance properties in simulated body fluid. Mater. Sci. Eng. C **69**, 675–684 (2016). https://doi.org/10.1016/j.msec.2016.07.057
17. G. Renaudin, S. Gomes, J.-M. Nedelec, First-row transition metal doping in calcium phosphate bioceramics: a detailed crystallographic study. Materials (Basel). **10**, 92 (2017). https://doi.org/10.3390/ma10010092
18. S. Gomes, A. Kaur, J.M. Grenèche, J.M. Nedelec, G. Renaudin, Atomic scale modeling of iron-doped biphasic calcium phosphate bioceramics. Acta Biomater. **50**, 78–88 (2017). https://doi.org/10.1016/j.actbio.2016.12.011
19. S. Khamseh, M. Ganjaee Sari, E. Alibakhshi, M. Nemati Valandaran, Hydrogen-free Cu: amorphous-C: N coating on TC4 titanium alloy: the role of gas ratio on mechanical and antibacterial potency. Prog. Color. Color. Coatings. **14**, 281–291 (2021). https://doi.org/10.30509/pccc.2021.166770.1102
20. A.V. Artem'ev, M.P. Davydova, X. Hei, M.I. Rakhmanova, D.G. Samsonenko, I.Y. Bagryanskaya, K.A. Brylev, V.P. Fedin, J.-S. Chen, M. Cotlet, J. Li, Family of robust and strongly luminescent CuI-based hybrid networks made of ionic and dative bonds. Chem. Mater. **32**, 10708–10718 (2020). https://doi.org/10.1021/ACS.CHEMMATER.0C03984
21. ASTM E9–09, Standard test methods of compression testing of metallic materials at room temperature (2018). https://www.astm.org/DATABASE.CART/HISTORICAL/E9-09R18.htm Accessed 16 Mar 2021
22. ASTM E384–17 Standard test method for microindentation hardness of materials (n.d.). https://www.astm.org/Standards/E384. Accessed 16 Mar 2021
23. K.O. Ighodalo, D. Obi, A. Agbogu, B.N. Ezealigo, A.C. Nwanya, S.L. Mammah, R. Bucher, M. Maaza, F.I. Ezema, The structural and optical properties of metallic doped copper (I) iodide thin films synthesized by SILAR method. Mater. Res. Bull. **94**, 528–536 (2017). https://doi.org/10.1016/j.materresbull.2017.06.032
24. L.A. DeSilva, J. Harwell, A. Gaquere-Parker, U.A.G. Perera, K. Tennakone, Thin films of copper (I) Iodide doped with iodine and thiocyanate. Phys. Status Solidi **214**, 1700520 (2017). https://doi.org/10.1002/pssa.201700520

25. M. Rabiei, A. Palevicius, A. Monshi, S. Nasiri, A. Vilkauskas, G. Janusas, Comparing methods for calculating nano crystal size of natural hydroxyapatite using x-ray diffraction. Nanomaterials **10**, 1627 (2020). https://doi.org/10.3390/nano10091627
26. M.H. Fathi, A. Hanifi, V. Mortazavi, Preparation and bioactivity evaluation of bone-like hydroxyapatite nanopowder. J. Mater. Process. Technol. **202**, 536–542 (2008). https://doi.org/10.1016/j.jmatprotec.2007.10.004
27. C.Y. Ooi, M. Hamdi, S. Ramesh, Properties of hydroxyapatite produced by annealing of bovine bone. Ceram. Int. **33**, 1171–1177 (2007). https://doi.org/10.1016/j.ceramint.2006.04.001
28. M. Rabiei, A. Palevicius, S. Nasiri, A. Dashti, A. Vilkauskas, G. Janusas, Relationship between young's modulus and planar density of unit cell, super cells (2 × 2 × 2), symmetry cells of Perovskite (caTiO3) lattice. Materials (Basel). **14**, 1258 (2021). https://doi.org/10.3390/ma14051258
29. M. Rabiei, A. Palevicius, A. Dashti, S. Nasiri, A. Monshi, A. Vilkauskas, G. Janusas, Measurement modulus of elasticity related to the atomic density of planes in unit cell of crystal lattices. Materials (Basel). **13**, 4380 (2020). https://doi.org/10.3390/ma13194380
30. T. Janusas, S. Urbaite, A. Palevicius, S. Nasiri, G. Janusas, Biologically compatible lead-free piezoelectric composite for acoustophoresis based particle manipulation techniques. Sensors. **21**, 483 (2021). https://doi.org/10.3390/s21020483
31. G. Renaudin, E. Jallot, J.M. Nedelec, Effect of strontium substitution on the composition and microstructure of sol-gel derived calcium phosphates. J. Sol-Gel Sci. Technol. 287–294 (2009). https://doi.org/10.1007/s10971-008-1854-5
32. S. Gomes, G. Renaudin, E. Jallot, J.M. Nedelec, Structural characterization and biological fluid interaction of sol-gel-derived Mg-substituted biphasic calcium phosphate ceramics. ACS Appl. Mater. Interfaces **1**, 505–513 (2009). https://doi.org/10.1021/am800162a
33. A. Doostmohammadi, A. Monshi, R. Salehi, M.H. Fathi, S. Karbasi, U. Pieles, A.U. Daniels, Preparation, chemistry and physical properties of bone-derived hydroxyapatite particles having a negative zeta potential. Mater. Chem. Phys. **132**, 446–452 (2012). https://doi.org/10.1016/j.matchemphys.2011.11.051
34. R.N. Panda, M.F. Hsieh, R.J. Chung, T.S. Chin, FTIR, XRD, SEM and solid state NMR investigations of carbonate-containing hydroxyapatite nano-particles synthesized by hydroxide-gel technique. J. Phys. Chem. Solids **64**, 193–199 (2003). https://doi.org/10.1016/S0022-3697(02)00257-3
35. V. Jokanović, D. Izvonar, M.D. Dramićanin, B. Jokanović, V. Živojinović, D. Marković, B. Dačić, Hydrothermal synthesis and nanostructure of carbonated calcium hydroxyapatite. J. Mater. Sci. Mater. Med. **17**, 539–546 (2006). https://doi.org/10.1007/s10856-006-8937-z
36. A. Wang, D. Liu, H. Yin, H. Wu, Y. Wada, M. Ren, T. Jiang, X. Cheng, Y. Xu, Size-controlled synthesis of hydroxyapatite nanorods by chemical precipitation in the presence of organic modifiers. Mater. Sci. Eng. C **27**, 865–869 (2007). https://doi.org/10.1016/j.msec.2006.10.001
37. R. Murugan, S. Ramakrishna, Bioresorbable composite bone paste using polysaccharide based nano hydroxyapatite. Biomaterials **25**, 3829–3835 (2004). https://doi.org/10.1016/j.biomaterials.2003.10.016
38. A. Karabulut, F.E. Baştan, G. Erdoɫan, F. Üstel, Heat treatment's effects on hydroxyapatite powders in water vapor and air atmosphere, in *AIP Conference Proceedings, American Institute of Physics Inc*, p. 020053 (2015). https://doi.org/10.1063/1.4914244
39. C.S. Chai, B. Ben-Nissan, Bioactive nanocrystalline sol-gel hydroxyapatite coatings. J. Mater. Sci. Mater. Med. **10**, 465–469 (1999). https://doi.org/10.1023/A:1008992807888
40. K. Tõnsuaadu, K.A. Gross, L. Pluduma, M. Veiderma, A review on the thermal stability of calcium apatites. J. Therm. Anal. Calorim. 647–659 (2012). https://doi.org/10.1007/s10973-011-1877-y
41. Mp. Rani, Preparation and characterization of cu doped hap for biomedical applications (n.d.)
42. M. Furko, V. Havasi, Z. Kónya, A. Grünewald, R. Detsch, A.R. Boccaccini, C. Balázsi, Development and characterization of multi-element doped hydroxyapatite bioceramic coatings on metallic implants for orthopedic applications. Bol. La Soc. Esp. Ceram. y Vidr. **57**, 55–65 (2018). https://doi.org/10.1016/j.bsecv.2017.09.003

43. M. Pretto, A.L. Costa, E. Landi, A. Tampieri, C. Galassi, Dispersing behavior of hydroxyapatite powders produced by wet-chemical synthesis. J. Am. Ceram. Soc. **86**, 1534–1539 (2003). https://doi.org/10.1111/j.1151-2916.2003.tb03510.x
44. P. Somasundaran, Y.H.C. Wang, Surface chemical characteristics and adsorption properties of apatite, in: Adsorpt. Surf. Chem. Hydroxyapatite, Springer US, pp. 129–149 (1984). https://doi.org/10.1007/978-1-4757-9012-2_9
45. L.M. Rodríguez-Lorenzo, M. Vallet-Regí, J.M.F. Ferreira, Colloidal processing of hydroxyapatite. Biomaterials **22**, 1847–1852 (2001). https://doi.org/10.1016/S0142-9612(00)00366-5
46. R.K. Mishra, S. Chander, D.W. Fuerstenau, Effect of ionic surfactants on the electrophoretic mobility of hydroxyapatite. Colloids Surf. **1**, 105–119 (1980). https://doi.org/10.1016/0166-6622(80)80040-0
47. ASTM D4187–82 Methods of test for zeta potential of colloids in water and waste water (Withdrawn 1990). https://www.astm.org/Standards/D4187.htm. Accessed 16 Mar 2021
48. P.M. Sirimanne, M. Rusop, T. Shirata, T. Soga, T. Jimbo, Characterization of transparent conducting CuI thin films prepared by pulsed laser deposition technique. Chem. Phys. Lett. **366**, 485–489 (2002). https://doi.org/10.1016/S0009-2614(02)01590-7
49. T. Supasai, S. Dangtip, P. Learngarunsri, N. Boonyopakorn, A. Wisitsoraat, S.K. Hodak, Influence of temperature annealing on optical properties of SrTiO 3 /BaTiO 3 multilayered films on indium tin oxide. Appl. Surf. Sci. **256**, 4462–4467 (2010). https://doi.org/10.1016/j.apsusc.2010.01.072
50. I.K. Vereshchagin, V.A. Nikitenko, S.G. Stoiukhin, Photoluminescence-excitation spectra of copper iodide single crystals, ZhPS. **36**, 848–851 (1982). https://ui.adsabs.harvard.edu/abs/1982ZhPS...36..848V/abstract. Accessed 16 Mar 2021
51. A. V. Artem'ev, E.A. Pritchina, M.I. Rakhmanova, N.P. Gritsan, I.Y. Bagryanskaya, S.F. Malysheva, N.A. Belogorlova, Alkyl-dependent self-assembly of the first red-emitting zwitterionic {Cu4I6} clusters from [alkyl-P(2-Py)3]+ salts and CuI: when size matters, Dalt. Trans. **48**, 2328–2337 (2019). https://doi.org/10.1039/C8DT04328K
52. D. Chen, Y. Wang, Z. Lin, J. Huang, X. Chen, D. Pan, F. Huang, Growth strategy and physical properties of the high mobility p-type cui crystal. Cryst. Growth Des. **10**, 2057–2060 (2010). https://doi.org/10.1021/cg100270d
53. W. Liu, Y. Fang, J. Li, Copper Iodide based hybrid phosphors for energy-efficient general lighting technologies. Adv. Funct. Mater. **28**, 1705593 (2018). https://doi.org/10.1002/ADFM.201705593
54. S. Yue, M. Gu, X. Liu, J. Zhang, S. Huang, B. Liu, C. Ni, Effect of ZnI2 cosolute on quality and performance of γ-CuI ultrafast scintillation crystal grown via evaporation method in acetonitrile solvent. Opt. Mater. (Amst) **66**, 308–313 (2017). https://doi.org/10.1016/j.optmat.2016.12.030
55. N.L. Ignjatović, L. Mančić, M. Vuković, Z. Stojanović, M.G. Nikolić, S. Škapin, S. Jovanović, L. Veselinović, V. Uskoković, S. Lazić, S. Marković, M.M. Lazarević, D.P. Uskoković, Rare-earth (Gd3+,Yb3+/Tm3+, Eu3+) co-doped hydroxyapatite as magnetic, up-conversion and down-conversion materials for multimodal imaging. Sci. Rep. **9** (2019). https://doi.org/10.1038/s41598-019-52885-0
56. M. Hosseinnezhad, J. Movahedi, S. Nasiri, High stability photosensitizers for dye-sensitized solar cells: synthesis, characterization and optical performance. Opt. Mater. (Amst). **109**, 110198 (2020). https://doi.org/10.1016/j.optmat.2020.110198
57. M. Hosseinnezhad, S. Nasiri, J. Movahedi, M. Ghahari, Improving the efficiency of organic sensitizers with various anchoring groups for solar energy applications. Sol. Energy **211**, 228–235 (2020). https://doi.org/10.1016/j.solener.2020.09.074
58. V.N. Ermakov, S.P. Kruchinin, T. Pruschke, J.K. Freericks, Thermoelectricity in tunneling nanostructures. Phys. Rev. B–Condens. Matter Mater. Phys. **92,** 155431 (2015). https://doi.org/10.1103/PhysRevB.92.155431.
59. R. Czerwieniec, M.J. Leitl, H.H.H. Homeier, H. Yersin, Cu(I) complexes–Thermally activated delayed fluorescence. Photophysical approach and material design, Coord. Chem. Rev. **325**, 2–28 (2016). https://doi.org/10.1016/j.ccr.2016.06.016

60. X. Hei, W. Liu, K. Zhu, S.J. Teat, S. Jensen, M. Li, D.M. O'Carroll, K. Wei, K. Tan, M. Cotlet, T. Thonhauser, J. Li, Blending Ionic and Coordinate bonds in hybrid semiconductor materials: a general approach toward robust and solution-processable covalent/coordinate network structures. J. Am. Chem. Soc. **142**, 4242–4253 (2020). https://doi.org/10.1021/JACS.9B13772
61. Z. Zheng, A. Liu, S. Wang, B. Huang, K.W. Wong, X. Zhang, S.K. Hark, W.M. Lau, Growth of highly oriented (110) γ-CuI film with sharp exciton band. J. Mater. Chem. **18**, 852–854 (2008). https://doi.org/10.1039/b719452h
62. X. Ji, W. Zhang, F. Ge, C. Wang, Y. Yin, K. Chen, Thermochromic behavior analysis of terminated polyurethane functionalized with rhodamine B derivative. Prog. Org. Coatings. **131**, 111–118 (2019). https://doi.org/10.1016/j.porgcoat.2019.02.022
63. H. Yin, Y. Li, J. Bai, M. Ma, J. Liu, Effect of calcinations temperature on the luminescence intensity and fluorescent lifetime of Tb3+-doped hydroxyapatite (Tb-HA) nanocrystallines. J. Mater. **3**, 144–149 (2017). https://doi.org/10.1016/j.jmat.2016.11.004
64. S.E. Derenzo, M.J. Weber, M.K. Klintenberg, Temperature dependence of the fast, near-band-edge scintillation from CuI, HgI2, PbI2, ZnO:Ga and CdS:In, in: Nucl. Instruments Methods Phys. Res. Sect. A Accel. Spectrometers, Detect. Assoc. Equip., North-Holland, pp. 214–219 (2002). https://doi.org/10.1016/S0168-9002(02)00705-2
65. N. Durairaj, S. Kalainathan, R. Kumar, Fluorescence emission and decay time studies on doped 1, 3, 5-triphenylbenzene scintillator crystal grown by solution growth technique fluorescence emission and decay time studies on doped 1, 3, 5-triphenylbenzene scintillator crystal grown by solution growth technique 37 (2017). https://doi.org/10.2412/mmse.4.45.791ï
66. B.N. Ezealigo, A.C. Nwanya, A. Simo, R.U. Osuji, R. Bucher, M. Maaza, F.I. Ezema, Optical and electrochemical capacitive properties of copper (I) iodide thin film deposited by SILAR method. Arab. J. Chem. **12**, 5380–5391 (2019). https://doi.org/10.1016/j.arabjc.2017.01.008
67. S. Kruchinin, T. Pruschke, Thermopower for a molecule with vibrational degrees of freedom. Phys. Lett. Sect. A Gen. At. Solid State Phys. **378**, 1157–1161 (2014). https://doi.org/10.1016/j.physleta.2014.01.054
68. Handbook of Solid-State Lasers–1st Edition. (n.d.). https://www.elsevier.com/books/handbook-of-solid-state-lasers/denker/978-0-85709-272-4. Accessed 16 Mar 2021
69. Semiconductors—Basic Data, Semicond—Basic Data. (1996). https://doi.org/10.1007/978-3-642-97675-9
70. A.N. Fletcher, Fluorescence emission band shift with wavelength of excitation. J. Phys. Chem. **72**, 2742–2749 (2002). https://doi.org/10.1021/J100854A009
71. M. Bakkal, A.J. Shih, S.B. McSpadden, C.T. Liu, R.O. Scattergood, Light emission, chip morphology, and burr formation in drilling the bulk metallic glass. Int. J. Mach. Tools Manuf **45**, 741–752 (2005). https://doi.org/10.1016/J.IJMACHTOOLS.2004.11.004
72. Auroral processes at the giant planets: energy deposition, emission mechanisms, morphology and spectra. Request PDF. (n.d.). https://www.researchgate.net/publication/302887929_Auroral_Processes_at_the_Giant_Planets_Energy_Deposition_Emission_Mechanisms_Morphology_and_Spectra. Accessed 29 Aug 2021
73. X. Luo, Y. Han, Z. Chen, Y. Li, G. Liang, X. Liu, T. Ding, C. Nie, M. Wang, F.N. Castellano, K. Wu, Mechanisms of triplet energy transfer across the inorganic nanocrystal/organic molecule interface. Nat. Commun. **111.11**, 1–10 (2020). https://doi.org/10.1038/s41467-019-13951-3
74. F.B. Dias, J. Santos, D.R. Graves, P. Data, R.S. Nobuyasu, M.A. Fox, A.S. Batsanov, T. Palmeira, M.N. Berberan-Santos, M.R. Bryce, A.P. Monkman, The role of local triplet excited states and D-A relative orientation in thermally activated delayed fluorescence: photophysics and devices. Adv. Sci. **3** (2016). https://doi.org/10.1002/advs.201600080
75. S. Nasiri, A. Dashti, M. Hosseinnezhad, M. Rabiei, A. Palevicius, A. Doustmohammadi, G. Janusas, Mechanochromic and thermally activated delayed fluorescence dyes obtained from D–A–D′ type, consisted of xanthen and carbazole derivatives as an emitter layer in organic light emitting diodes. Chem. Eng. J. 131877 (2021). https://doi.org/10.1016/J.CEJ.2021.131877
76. H.D. Burrows, L.S. Ernestova, T.J. Kemp, Y.I. Skurlatov, A.P. Purmal, A.N. Yermakov, Kinetics and mechanism of photodegradation of chlorophenols **23**, 145–207 (2019). https://doi.org/10.3184/007967498103165022

77. L. Arnaut, S. Formosinho, H. Burrows, Transitions between electronic states, chem. Kinet, 385–435 (2007). https://doi.org/10.1016/B978-044452186-6/50015-8
78. T.Y. Chu, O.K. Song, Hole mobility of N, N′ -bis(naphthalen-1-yl)- N, N′ -bis(phenyl) benzidine investigated by using space-charge-limited currents. Appl. Phys. Lett. **90**, 203512 (2007). https://doi.org/10.1063/1.2741055
79. K. Prasad, O. Bazaka, M. Chua, M. Rochford, L. Fedrick, J. Spoor, R. Symes, M. Tieppo, C. Collins, A. Cao, D. Markwell, K. (Ken) Ostrikov, K. Bazaka, Metallic biomaterials: current challenges and opportunities. Materials (Basel). **10**, 884 (2017). https://doi.org/10.3390/ma10080884
80. W. Luo, C. Zeng, X. Du, C. Leng, W. Yao, H. Shi, X. Wei, C. Du, S. Lu, Copper thiocyanate/copper iodide based hole transport composites with balanced properties for efficient polymer light-emitting diodes. J. Mater. Chem. C. **6**, 4895–4902 (2018). https://doi.org/10.1039/c7tc04842d
81. D.S. Gomes, A.M.C. Santos, G.A. Neves, R.R. Menezes, A brief review on hydroxyapatite production and use in biomedicine. Ceramica. **65**, 282–302 (2019). https://doi.org/10.1590/0366-69132019653742706
82. M. Schuth, F. Vössing, L. Yang, A shearographic endoscope for nondestructive test. J. Hologr. Speckle. **1**, 46–52 (2004). https://doi.org/10.1166/jhs.2004.007
83. Y.M. He, C.J. Tay, H.M. Shang, A new method for generating and analysing digital speckle shearing correlation fringe patterns. Opt. Laser Technol. **30**, 27–31 (1998). https://doi.org/10.1016/S0030-3992(97)00062-5

Index

M. Rabiei et al., *Hydroxyapatite-Based Nanocomposites*, Springer Series in Chemical Physics 127, https://doi.org/10.1007/978-3-032-12583-5

MIX
Papier aus verantwortungsvollen Quellen
Paper from responsible sources
FSC® C105338

FSC
www.fsc.org

If you have any concerns about our products,
you can contact us on
ProductSafety@springernature.com

In case Publisher is established outside the EU,
the EU authorized representative is:
Springer Nature Customer Service Center GmbH
Europaplatz 3, 69115 Heidelberg, Germany

Printed by Libri Plureos GmbH
in Hamburg, Germany